280 Current Topics in Microbiology and Immunology

Springer-Verlag Berlin Heidelberg GmbH

E. Heber-Katz (Ed.)

Regeneration: Stem Cells and Beyond

With 42 Figures and 11 Tables

Springer

Ellen Heber-Katz, Ph.D.
The Wistar Institute
3601 Spruce Street
Philadelphia, PA 19104
USA
e-mail: heberkatz@mail.wistar.upenn.edu

Cover Illustration by James Hayden, RBP and Doug Thayer:
The MRL mouse ear hole (upper inset) completely closes without scarring within 30 days (lower inset). An excised ear tissue specimen is shown with abundant hair and follicles from a mouse ear unstained and visualized with autofluorescence.

ISSN 0070-217X
ISBN 978-3-642-62321-9 ISBN 978-3-642-18846-6 (eBook)
DOI 10.1007/978-3-642-18846-6

http://www.springer.de

Originally published by Springer-Verlag Berlin Heidelberg New York in 2004
Softcover reprint of the hardcover 1st edition 2004
Library of Congress Catalog Card Number 15-12910

Cover Design: *design & production GmbH, Heidelberg*
Typesetting: Stürtz AG, Würzburg
Production Editor: Angélique Gcouta, Berlin
Printed on acid-free paper SPIN: 10914922 27/3020 5 4 3 2 1 0

Preface

The word "regeneration" evokes both strange and wondrous images. This includes not only the sea-shore starfish whose severed arm quickly grows back, or the regrowth of limbs on the tails of newts, but most directly it relates to modern medicine with the image of Prometheus having his liver being eaten every night and having it grow back during the day. Perhaps the ancient Greeks saw the same starfish on Mediterranean shores and developed their myths from that. Or perhaps pharaonic physicians saw healing of old battle wounds at autopsy and noticed regenerating liver. It is of interest that sacrificial entrails focused on the liver as pleasing to the gods. Did those priests observe regeneration? This is all lost to history but even today the regenerating limb inspires delight.

At the present time, of course, with the modern molecular techniques for cell manipulation and cloning, we can truly imagine the possibility of self-renewal, and if not total, then at least in terms of organs and limbs. It is thus no surprise that there should be a focus on the major self-renewing population of cells, the stem cells. The general understanding is that this is a population of cells that is maintained through development and into adulthood, that this population does not mature, but rather divides and provides a continual population of multipotent cells. The hope is that these cells, using the appropriate inducing factors, can generate mature cells in the proper milieu in a controlled manner.

This, however, is not the only process for such self-renewal. There is another possible route to tissue and organ restoration which is through cells that are already mature and that can de-differentiate into a less mature and more plastic cell type. It has been shown that such cells can re-differentiate into the same or a different mature population of cells. This pathway has been confirmed in the newt regeneration model and can account for all regeneration seen. The newt blastema formed after injury is the probable source of such cells and these cells might present as a type of fibroblast. In this new world of trans-differentiation and cytoplasm-directed reprogramming, issues of DNA and histone modifications and remodeling are becoming a major scientific thrust in regenerative studies.

While evolution and phylogeny are wholly devoted to a comparative study of species, perhaps no other area of biology shows the range of species variability as the capacity for regeneration. And, while the general pattern is clear, that regenerative capacity declines through phylogeny, there are loads of surprises. Some flatworm species are the superstars of regeneration while others are bereft of the trait. Adult axolotls are paradigmatic healers while adult *Xenopus* are as incapable of regeneration as typical mammals. The mouse parallels the human in that only liver and the tips of digits show full organ regeneration except for one strain, the MRL mouse, which can heal heart tissue as well as the zebrafish. Implicit in the study of regeneration is cross-species comparison. Indeed, many believe that the critical pathways for regeneration can only be deeply understood in this context.

In this volume of *Current Topics in Microbiology and Immunology* entitled "Regeneration: Stem Cells and Beyond," we present a unique range of examples of the regenerative response and the role of stem cells from the amphibian to human. In the first chapter by David Stocum, the issue of blastema formation and patterning is described in the amphibian, which includes three classes: (1) urodeles (i.e., salamanders, newts, axolotls); (2) anurans (i.e., frogs and toads); and (3) apodans (e.g., legless amphibians). This group of animals contains the best known and most classic models of regeneration. Dr. Stocum discusses the key aspects of blastema formation, de-differentiation and re-differentiation, and the importance of patterning in limb regeneration. While not excluding a role for stem cells, Stocum and others make a strong case for the phenomenon of de-differentiation and re-differentiation for the formation of new tissue.

In contrast to this, Henry Young presents an alternative view by identifying reserve stem cells in the amphibian. Furthermore, he has extended this to almost all human tissues. He was an early proponent of this notion and presents data showing resident stem cells in adult tissues of many mammals.

In the third chapter, there is an example of one of the first noted mammalian tissues to show a regenerative response, the skin, and more specifically the epidermis. Epidermal stem cells and their properties in culture are discussed. This system has allowed Natalia Popova and Rebecca Morris to examine stem cells in culture and analyze their proliferative, differentiative, and clonal potential as well as to examine the genetics of such properties. Such potential fits well with the in vivo characteristics identified.

In contrast, Bruce Rosengard and colleagues discuss an example of one organ in which there has been little or no evidence of self-renewal in mammals—the heart. They discuss the future of using tissue engineering approaches to address healing in this critical organ.

Our laboratory has studied the unique regeneration potential of the MRL strain of mouse. We first noted that through-and-through ear hole punches used for identification closed scarlessly and showed regeneration of cartilage and hair follicles. This model has been very fruitful, especially as the MRL can, in fact, regenerate large portions of ablated heart tissue.

Finally, it now seems that all roads lead to cells that have the plasticity to become something else, not only in the amphibian but also in the mammal. And we should be willing to accommodate any or all cells, whether they remain as stem cells from birth or have matured and can then re-differentiate.

Philadelphia, 2003 Ellen Heber-Katz

List of Contents

Amphibian Regeneration and Stem Cells
D.L. Stocum 1

Existence of Reserve Quiescent Stem Cells in Adults, From Amphibians to Humans
H.E. Young 71

Genetic Regulation of Mouse Stem Cells: Identification of Two Keratinocyte Stem Cell Regulatory Loci
N.V. Popova and R.J. Morris 111

Myocardial Tissue Engineering and Regeneration as a Therapeutic Alternative to Transplantation
A.S. Krupnick, D. Kreisel, M. Riha, K.R. Balsara, and B.R. Rosengard 139

Spallanzani's Mouse: A Model of Restoration and Regeneration
E. Heber-Katz, J.M. Leferovich, K. Bedelbaeva, and D. Gourevitch . . . 165

Subject Index 191

List of Contributors

(Their addresses can be found at the beginning of their respective chapters.)

Balsara, K.R. 139

Bedelbaeva, K. 165

Gourevitch, D. 165

Heber-Katz, E. 165

Kreisel, D. 139

Krupnick, A.S. 139

Leferovich, J.M. 165

Morris, R.J. 111

Popova, N.V. 111

Riha, M. 139

Rosengard, B.R. 139

Stocum, D.L. 1

Young, H.E. 71

Amphibian Regeneration and Stem Cells

D. L. Stocum

Department of Biology and the Indiana University Center for Regenerative Biology and Medicine, School of Science, Indiana University-Purdue University Indianapolis, 402 N. Blackford St., Indianapolis, IN 46202, USA
E-mail: dstocum@iupui.edu

1 Introduction 2
2 Events of Limb Regeneration 5
3 Mechanism of Blastema Formation 8
3.1 Degradation of the ECM 9
3.2 Re-entry of Myonuclei into the Cell Cycle 10
3.3 Loss of Phenotypic Specialization 11
4 Requirements for Proliferation of Blastema Cells 14
4.1 AEC Factors 14
4.2 Neural Factors 19
4.3 Do the AEC and Nerves Have the Same or Different Functions? 22
5 Developmental Plasticity of Dedifferentiated Cells 25
5.1 Plasticity of Adult Mammalian Reserve Stem Cells 25
5.2 Plasticity of Blastema Cells 26
5.3 Do Reserve Stem Cells Contribute to the Blastema? 27
6 Spatial Organization of Tissue Patterns in the Regenerating Limb 28
6.1 The Blastema is a Self-Organizing System 28
6.2 A Mechanism of Self-Organization Based on Inherited Positional Boundaries and Intercalary Regeneration 30
6.3 The Use of Retinoids to Investigate the Physical Basis of Positional Identity 34
6.3.1 Nuclear Receptors and Cytoplasmic Binding Proteins Mediate the Biological Effects of Retinoids 37
6.3.2 The δ2 RAR Mediates Proximalization of Positional Identity 38
6.4 Transcription Factors and Signaling Molecules Involved in Specifying Positional Identity 40
6.4.1 Proximodistal Axis 41
6.4.2 Anteroposterior Axis 43
6.4.3 Dorsoventral Axis 44
7 Other Regenerating Systems 45
7.1 Neural Retina Regeneration 46
7.2 Lens Regeneration 47
8 Relevance of Amphibian Regeneration to Regenerative Medicine 49
9 Summary and Conclusions 52
References 53

Abstract Larval and adult urodeles and anuran tadpoles readily regenerate their limbs via a process of histolysis and dedifferentiation of mature cells local to the amputation surface that accumulate under the wound epithelium as a blastema of stem cells. These stem cells require growth and trophic factors from the apical epidermal cap (AEC) and the nerves that re-innervate the blastema for their survival and proliferation. Members of the fibroblast growth factor (FGF) family synthesized by both AEC and nerves, and glial growth factor, substance P, and transferrin of nerves are suspected survival and proliferation factors. Stem cells derived from fibroblasts and muscle cells can transdifferentiate into other cell types during regeneration. The regeneration blastema is a self-organizing system based on positional information inherited from parent limb cells. Retinoids, which act through nuclear receptors, have been used in conjunction with assays for cell adhesivity to show that positional identity of blastema cells is encoded in the cell surface. These molecules are involved in the cell–cell signaling network that re-establishes the original structural pattern of the limb. Other systems of interest that regenerate by histolysis and dedifferentiation of pigmented epithelial cells are the neural retina and lens. Members of the FGF family are also important to the regeneration of these structures. The mechanism of amphibian regeneration by dedifferentiation is of importance to the development of a regenerative medicine, since understanding this mechanism may offer insights into how we might chemically induce the regeneration of mammalian tissues.

1 Introduction

Regeneration is a process essential to life. All organisms fight an uphill thermodynamic battle against entropic processes that tend to disorganize their structure and thus decrease their ability to function (do work). The entropic processes eventually win, resulting in the death of the individual. To avoid death of the species, individuals funnel energy into the generation of new individuals, by asexual or sexual reproduction. Asexual reproduction produces clones; for example the repeated division of single celled organisms in good nutritional times, or the budding off by multicellular organisms, such as hydra, of a younger version of themselves. Sexual reproduction produces nonclones by recombining the genomes of sperm and eggs.

While death of the individual is ultimately unavoidable, perpetuation of the species does require that a significant proportion of the individuals comprising a population avoid premature death in order to reach reproductive maturity and beyond. This requires that each organism use the bulk of its energy to grow and maintain its integrity by adding to its structure and constantly regenerating worn-out or damaged parts. Richard J. Goss has concisely summed up the relationship between regeneration, life and death in the following words. 'If there were no regeneration there could be no life. If everything regenerated there would be no death. All organisms exist between these two extremes. Other things being equal, they tend toward the latter end of the spectrum, never quite achieving immortality because this would be incompatible with reproduction' (Goss 1969).

Within individual organisms, regeneration takes place on several levels of biological organization. On a cellular level, molecules and organelles turn over and must be replaced. Cells can adjust the balance of protein synthesis and degradation in response to biochemical or mechanical load. For example, cardiomyocytes replace most of their molecules over the course of 2 weeks and adjust their rate of protein synthesis upwards under a sustained increase in blood pressure, becoming hypertrophied (Gevers 1984). Neurons of the peripheral nervous system can regenerate axons, using the Schwann cell nerve tubes as guides (Yannas 2001). The free-living unicellular protozoans can regenerate complete cells after removal of large fragments as long as nuclear material is present in the remaining part (Goss 1969). For example, as little as one-eightieth of an amoeba is capable of reconstituting a complete amoeba (Vorontsova and Liosner 1960).

Tissue regeneration in multicellular organisms is accomplished by mitotically-competent cells residing within tissues. In some cases these cells are fully differentiated. Thus hepatocytes of the vertebrate liver respond to partial hepatectomy by mitosis while maintaining all their differentiated functions (Michaelopoulos and DeFrances 1997). This mode of regeneration is called compensatory hyperplasia. The tissues of most multicellular organisms, however, are regenerated by undifferentiated stem cells. In the majority of cases, these are reserve stem cells, suspended in an early state of differentiation as cells around them continue to differentiate during embryonic development. These stem cells may have unipotent to multipotent developmental potential. Reserve stem cells have long been recognized as the source of renewal and regeneration for vertebrate tissues such as blood, immune cells, epithelia, bone, muscle, and olfactory bulb and nerve (Stocum 2001). In some species, such as planaria and hydra, they can regenerate whole organisms from fragments of the body (Goss 1969; Alvarado 2000). In vertebrates, reserve stem cells, along with the more than 200 differentiated cell types of the body, arise from a small number of pluripotent embryonic stem cells (ESCs) that comprise the early embryo (Marshak et al. 2001).

The pluripotency and unlimited growth potential of cultured human embryonic stem cells and the recently demonstrated plasticity of reserve adult stem cells has given rise to an exciting vision of 'regenerative medicine' in which these cells or their derivatives are transplanted to replace tissues damaged by injury or disease, or are induced in vivo to regenerate the tissues (Weissman 2000a; Orkin and Morrison 2002). However, there is much that we must learn yet about the biology of regeneration in order to develop such therapies. Thus it is more appropriate to call this new field 'regenerative biology and medicine' to emphasize that understanding the biology of regeneration is prerequisite for establishing a regenerative medicine.

Different species of vertebrates vary in their ability to regenerate, the anuran (frogs and toads) and urodelan (salamanders and newts) amphibians having the greatest regenerative capacity. Amphibians use compensatory hyperplasia and reserve stem cells to regenerate all of the tissues regenerated by other vertebrates. What sets them apart from the rest is their ability to regenerate complex structures by dedifferentiation (Stocum 1995; Mescher 1996; Brockes 1997). Frog and toad tadpoles can regenerate tails, limb buds, lens, neural retina, and spinal cord. Larval urodeles regenerate tails, limbs, jaws, and spinal cord, but not lens or neural retina. Larval and adult newts are perhaps the best regenerators of all, being able to replace all of these structures, plus intestine and heart muscle. The spinal cord is regenerated by an ependymal cell response that bridges the lesion, provides a favorable environment for axon regrowth and, in some cases, regenerates new neurons (Egar and Singer 1981; Stensaas 1983; Chernoff 1996). Cardiac muscle cells divide after injury with their contractile fibrils in a partially intact state (Oberpriller and Oberpriller 1991). All the other structures mentioned are regenerated by stem cells derived via the dedifferentiation of mature cells at the site of injury. Dedifferentiation is a unique method of producing stem cells, seen otherwise only in fish, which regenerate fins, and lizards, which regenerate tails. Regenerated lizard tails, however, lack vertebrae, spinal cord, and spinal ganglia and consist only of a cartilaginous rod and undifferentiated ependymal tube. Because the amphibians exhibit all of the known mechanisms of regeneration, they are an important experimental tool for expanding our understanding of regeneration and how to apply this knowledge to induce the regeneration of nonregenerating mammalian tissues and complex structures.

The primary focus of this paper will be on amphibian limb regeneration, with secondary references to lens, neural retina, and other structures regenerated by dedifferentiation. The limb is a complex structure that regenerates well in early frog and toad tadpoles, larval salamanders, and adult newts. Its tissues are easy to manipulate surgically, culture systems for mature and dedifferentiated cells are available (Stocum 2000), and the molecular biology of limb regeneration is sufficiently advanced, or soon will be, to enable characterization of the gene activity that drives regeneration (Geraudie and Ferretti 1998).

2
Events of Limb Regeneration

The regeneration of an amputated urodele or anuran tadpole limb goes through a typical series of morphological and histological stages (Iten and Bryant 1973; Smith et al. 1974; Tank et al. 1976; Stocum 1979; Young et al. 1983). The wound is closed by basal epidermal cells migrating from the cut edges of the skin. Undifferentiated cells accumulate under the wound epidermis to form a regeneration blastema, while at the same time, the apical wound epidermis thickens to form the AEC. Whether formation of the AEC is a process autonomous to the wound epidermal cells or is induced by the underlying blastema cells is not known. The AEC is the homolog of the apical ectodermal ridge (AER) of the embryonic limb bud in anuran and amniote embryos. Both the AER and the AEC are essential for the outgrowth of the limb bud or blastema. Urodele limb buds do not have a morphologically recognizable AER, but the distal ectoderm functions like one in promoting the outgrowth of the limb bud mesenchyme (Stocum 1984). The outer layers of the AEC are protective, whereas its basal layers appear to be anatomically and functionally equivalent to the AER of amniote embryonic limb buds (Christensen and Tassava 2000). The cut ends of the nerves innervating the limb begin sprouting into the tip of the amputated limb within a few days of amputation. The sensory fibers innervate the wound epidermis,

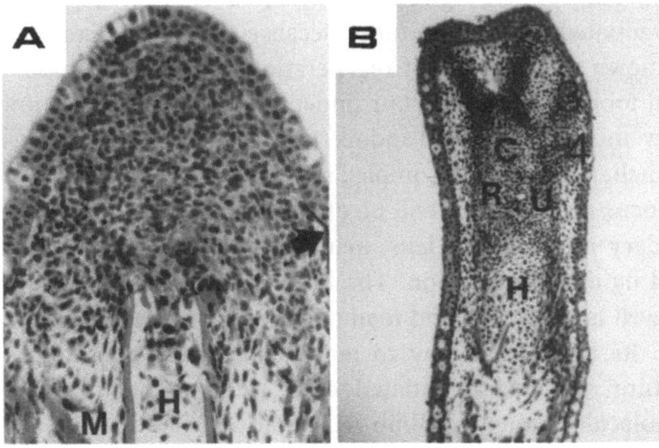

Fig. 1A, B Two stages of forelimb regeneration in larval *Ambystoma maculatum*. The limb was amputated through the mid-humerus. **A** Medium bud stage blastema. The wound epidermis is thickened to form the AEC. The *arrow* points to the basement membrane of the skin. The basement membrane is being re-constituted under the epidermis in the proximal part of the blastema, but is absent from the distal part of the blastema. *H*, humerus; *M*, muscle of the limb stump. **B** Two-fingerbud stage. The blastema cells are differentiating the amputated segments of the limb. *H*, humerus; *R*, radius; *U*, ulna; *C*, carpal mass; *3*, *4*, the third and fourth digits that are not yet visible externally

while the motor fibers ramify throughout the blastema cells, where they will eventually re-innervate the new muscles (Van Stone 1955; Piatt 1957).

Once a blastema has formed, regeneration seems to largely recapitulate the events of limb ontogenesis (Fig. 1). The undifferentiated cells divide by mitosis, enlarging the blastema into a conical bud. With further growth of the blastema, its cells withdraw from the cell cycle and differentiate in the histogenetic and morphogenetic patterns of the amputated parts. As muscles differentiate, they are re-innervated by regenerating motor nerve fibers. Differentiation and morphogenesis take place in a proximal-to-distal and anterior-to-posterior sequence, except that in the proximodistal (PD) axis, the digits appear to begin differentiation prior to the carpals or tarsals. In the anteroposterior (AP) axis, the radius differentiates faster than the ulna, and the anterior-most two digits appear first, followed by the remaining two (hand) or three (foot) digits. This anterior-to-posterior sequence of differentiation is unique to urodeles;

the regenerated and embryonic limb buds of anuran amphibians and the limb buds of amniotes follow a posterior-to-anterior sequence (Shubin and Alberch 1986). Differentiation along the dorsoventral (DV) axis appears to be simultaneous. The remainder of the regenerative process consists of growth to match the size of the unamputated limb.

The forelimb buds of urodele larvae first appear at 7.5 days post-fertilization and are fully differentiated 3 weeks later. The larvae do not metamorphose until 2–3 months after this. The hindlimb buds of anuran tadpoles first appear at 4 days post-fertilization and are fully differentiated between 5 and 6 weeks later, as the tadpole begins to undergo metamorphosis. The undifferentiated hindlimb buds of anuran tadpoles regenerate well, but lose the capacity for regeneration in a proximal-to-distal sequence in concert with differentiation of the limb segments (Guyenot 1927; Dent 1962). Loss of regenerative capacity occurs on time schedules and to degrees that vary in different groups of anurans (Stocum 1995). In the first group, typified by Ranid tadpoles, regeneration declines to zero prior to the onset of metamorphosis and amputated froglet limbs simply scar over. In the second group, exemplified by the toads *Alytes obstetricans* and *Bufo regularis*, regeneration also declines to zero, but a bit later, during metamorphosis. The third group consists of species in which regenerative capacity declines during metamorphosis, but not to zero. These include some of the Pipidae, Discoglossidae and Microhylidae. The amputated limbs of froglets/toadlets and adults are capable of regeneration, but the regenerates are hypomorphic, ranging from spike-like to spatulate outgrowths of cartilage and connective tissue, sometimes with rudimentary digits. *Xenopus laevis* is the best-studied example of this group.

The hindlimb buds of *Xenopus* begin to differentiate proximally at stage 51 (Nieuwkoop and Faber 1967) when the mesenchymal condensation of the femur appears. By stage 52, the femur is procartilaginous, upper leg muscle condensations have appeared and the mesenchymal condensations of the tibia and fibula have emerged. The femur begins to chondrify at stage 53, the tibia and fibula are procartilaginous, and the mesenchymal condensations of the foot are present. Chondrification, ossification and muscle differentiation sweep proximally to distally until differentiation is complete by stages 59–60. Up until stage 53 amputation through the femur results in the formation of a normal blastema and complete limb regeneration. At stage 53, regenerative capacity begins to decline, starting in the femur region. Amputation through the femur at this stage results in regenerates with malformations or missing structures, but amputation through the tibia–fibula or ankle region still results in normal regeneration. Thereafter, regeneration becomes increasingly hypomorphic in a proximal-to-distal progression until finally, as the animal progresses through metamorphosis, only hypomorphic spikes of cartilage regenerate from any level of amputation (Dent 1962; Overton 1963; Korneluk and Liversage 1984; Muneoka et al. 1986). The quality of regeneration is less after amputation through regions of ossification than through the joints of the limb (Wolfe et al. 2000).

The increasing hypomorphism of limb regeneration with advancing developmental stage in *Xenopus* is correlated with changes in the cellular features of the blastema (Korneluk and Liversage 1984; Wolfe et al. 2000). Histolysis is minimal and the blastema cells that appear are fibroblastic in morphology, rather than mesenchymal. There is an increase in basement membrane and dermal tissue invasion under the wound epidermis, AEC thickness declines, and there is decreased ability to support the sprouting of nerves and blood vessels into the blastema (Wolfe et al. 2000). The fibroblastic cells proliferate and then differentiate into several cartilage nodules, which fuse to form a symmetrical cartilage spike of varying length. The spike is surrounded by connective tissue, but no muscle.

The loss of regenerative capacity in *Xenopus* limbs appears to be the result of intrinsic changes in limb cells as they differentiate, not to systemic factors. Transplantation of regeneration-deficient limb buds to regeneration-competent limb buds and vice versa does not alter the regenerative capacity of the donor limb bud (Sessions and Bryant 1988; Filoni et al. 1991).

3
Mechanism of Blastema Formation

The extracellular matrix (ECM) of the tissues local to the amputation surface is degraded for a short distance proximal to the amputation plane, resulting in the loss of tissue organization (histolysis), liberation of cells from their tissue organization, and dedifferentiation of the liberated cells (Thornton 1968; Stocum 1995). Light and electron microscopic studies had indicated that blastema cells arise by the dedifferentiation of chondrocytes, muscle cells, dermal fibroblasts and Schwann cells (Thornton 1938, 1968; Hay 1959). Definitive proof of dedifferentiation, however, required the tracking of marked differentiated cells into the regeneration blastema. Triploid cartilage and dermis have been grafted to diploid limbs of axolotls (Steen 1968; Muneoka et al. 1986). Mutant 1-nu myofibers have been implanted into 2-nu *Xenopus* hindlimbs (Steen 1973), and myotubes labeled in vitro with fluorescent rhodamine-conjugated lysinated dextran (Lo et al. 1993), the tracker dye PKH26, or a retrovirus expressing alkaline phosphatase (Kumar et al. 2000), have been implanted into regenerating adult newt limbs. In each case, labeled undifferentiated cells appeared in the blastema, demonstrating that fibroblasts, chondrocytes, and muscle cells all undergo dedifferentiation. In addition, endogenous myofibers of the axolotl tail labeled with rhodamine dextran in vivo have been shown directly to dedifferentiate and contribute to the blastema after amputation (Echeverri et al. 2001).

Dedifferentiation involves several processes. Liberated cells re-enter the cell cycle, lose their phenotypic specializations, and change their pattern of gene activity to replace the adult matrix with one that is more like that of the limb bud (Stocum 1995). Regardless of their origin, the morphological appearance of all the blastema cells is that of an embryonic-like mesenchymal cell, similar to that of the limb bud. These cells can be considered stem cells. They can differentiate either into the cell of origin or into some other cell type (see Sect. 5) and they are renewable, not in the conventional sense of renewal by asymmetric division, but in their ability to be produced anew from differentiated cells in successive rounds of regeneration. Re-entry into the cell cycle and the loss of phenotypic specialization have so far been studied only in cultured newt muscle cells. Myofibers are structurally the most complex cells that contribute to the blastema. Not only must they re-enter the cell cycle, shut down the production of myogenic regulatory proteins, and dismantle the contractile protein apparatus, they must also fragment into mononucleate cells.

3.1 Degradation of the ECM

The ECM is degraded by acid hydrolases such as cathepsin D, acid phosphatase (Ju and Kim 1998), β-glucuronidase, and carboxylic ester hydrolases (Schmidt 1966), and by matrix metalloproteinases (MMPs) (Grillo et al. 1968; Dresden and Gross 1970). So far, newt and axolotl MMPs 2 and 9 (gelatinases) and MMP3/10a and b (stromelysins) have been cloned and their expression shown to be elevated during the period of blastema formation (Miyazaki et al. 1996; Ju and Kim 1998; Park and Kim 1999; Yang et al. 1999). Acid hydrolase and MMP activity are sharply increased in the regenerating limbs of animals treated with retinoic acid (RA), which causes more extensive dedifferentiation than normal (Ju and Kim 1994; Park and Kim 1999).

Acid hydrolases are undoubtedly released from injured and dying cells, and are particularly active in chondroclasts and osteoclasts, which degrade cartilage and bone matrix (Teitelbaum 2000) and are abundant in regenerating limbs (Stocum 1979). The spatial patterns of expression of MMPs have not been analyzed, so it is unclear which cell types produce these proteases. Potential candidates are macrophages, wound epidermis, neurites extending into the blastema and the blastema stem cells themselves. During embryogenesis, extending neurites cut a path through the ECM via their production of MMPs (Pittman and Buettner 1989). Epidermal cells also secrete MMPs to cut through the plasma clot during re-epithelialization of a wound, and macrophages produce MMPs to degrade disorganized matrix within a wound (Clark 1996).

Matrix degradation ceases as the cells of the blastema undergo differentiation and morphogenesis, and re-synthesize the mature ECM. Cessation of matrix degradation is likely to involve the upregulation of tissue inhibitors of metalloproteinases (TIMPS) as well as the downregulation of MMPs and acid hydrolases at any or all levels of gene regulation, from transcription to protein modification. Neither the regulation of protease activity nor expression of TIMPS have been systematically analyzed in regenerating limbs, however.

3.2 Re-entry of Myonuclei into the Cell Cycle

The nuclei of cultured newt and mouse myoblasts are induced to re-enter the cell cycle by either serum or the several growth factors in serum which act as mitogens: platelet-derived growth factor (PDGF), fibroblast growth factor-2 (FGF-2), epidermal growth factor (EGF), insulin-like growth factor-I (IGF-I), transforming growth factor β, and tumor necrosis factor α (Tanaka et al. 1997). These growth factors stimulate phosphorylation of the retinoblastoma (Rb) protein, releasing it from E2F transcription factors, which then activate genes whose products are essential for DNA synthesis. Serum also stimulates DNA synthesis by the nuclei of cultured newt myotubes, but not the nuclei of cultured C2C12 mouse myotubes. However, the same serum growth factors known to stimulate DNA synthesis in newt and mouse myoblasts fail to do so in newt myotubes (Tanaka et al. 1997). These results suggest that, as they differentiate from myoblasts, newt myotubes become refractory to these growth factors, but express a receptor that can respond to another protein in serum that stimulates DNA synthesis. Mouse myotubes appear to lack a functional receptor for this protein. The protein has not been identified, but experiments in which serum fractions were tested on cultured myoblasts and myotubes indicate that it requires thrombin for activation. Flurogenic assays have shown that thrombin activity increases during blastema formation (Tanaka et al. 1999).

Re-entry into the cell cycle does not require fragmentation of the multinucleated myotubes into mononucleate cells. This was shown by an experiment in which cultured newt myotubes were labeled with fluorescein dextran and X-irradiated or transfected with the p16 gene for cyclin-dependent kinase 4/6 inhibitor to block entry into S phase. These myotubes were then co-implanted with control myotubes labeled with Texas Red–Dextran into limb blastemas. Both the control and cell cycle-inhibited myotubes fragmented into mononucleate cells, but only the control myotubes synthesized DNA and divided, as shown by BrdU incorporation (Velloso et al. 2000). Completion of mitosis does appear to require mononucleate cell status. The mechanism of fragmentation of myotubes into single cells is not yet known, nor is it known whether the thrombin-activated protein is necessary to drive mononucleate cells such as chondrocytes and fibroblasts into the cell cycle as well, or whether this is a feature unique to myofibers.

3.3 Loss of Phenotypic Specialization

We do not have a clear understanding of the mechanisms that initiate the loss of phenotypic specialization during limb regeneration. Degradation of the ECM by proteases would break contacts between ECM molecules and integrin receptors, leading to changes in cell shape and reorganization of the actin cytoskeleton. This reorganization in turn might activate signal transduction pathways that induce the upregulation of enzymes which dismantle the phenotype-specific internal structure of the cells.

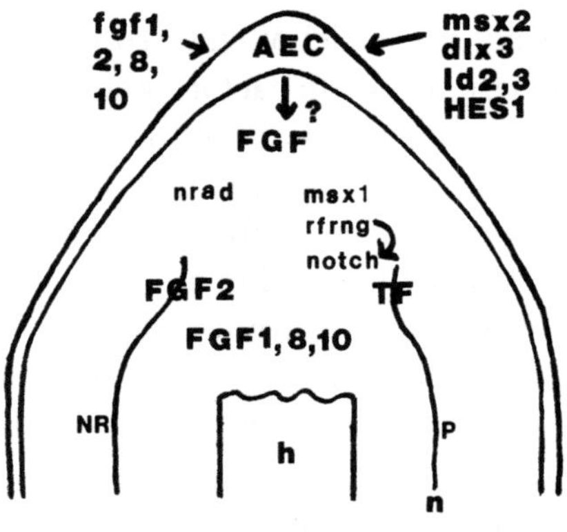

Fig. 2 Genes expressed by the AEC, nerves, and blastema cells that may be involved in dedifferentiation and proliferation. Growth factor genes expressed by the AEC are *fgf-1, 2, 8,* and *10.* AEC transcription factor genes are *msx-2, dlx-3, Id 2* and *3,* and *HES 1.* Whether FGFs are secreted into the blastema by the AEC is unknown, as indicated by the *arrow* and *question mark.* Growth factors associated with nerves (*n*) and nerve cell bodies are neuregulin (*NR*), substance P (*P*), FGF-2, and transferrin (*TF*): The blastema stem cells express FGF-1, 8, and 10, the intracellular signaling molecule nrad, the transcription factor msx1, and rfrng and notch of the notch receptor pathway. *h,* humerus; *n,* nerve. Neuregulin and substance P are depicted as being in the nerve fibers, but actually have been demonstrated only in the nerve cell bodies of the spinal ganglia

What these pathways are is unknown, but several genes have been identified whose products might be involved in the initiation or maintenance of dedifferentiation (Fig. 2). The transcription factor msx1 is expressed in undifferentiated limb bud mesenchyme and regeneration blastemas (Hill et al 1989; Robert et al. 1989; Crews et al. 1995; Simon et al. 1995). The *msx1* gene inhibits myogenesis by downregulating MyoD when ectopically expressed in mouse myoblasts in vitro (Song et al. 1992; Woloshin et al. 1995). C2C12 mouse myotubes fragment into mononucleate cells and undergo morphological and molecular dedifferentiation when transfected in vitro with *msx1* (Odelberg et al. 2001). Interestingly, these mononucleate cells behave like mesenchymal stem cells of the bone marrow stroma (Pittenger et al. 1999), being able to differentiate in vitro into cells expressing chondrogenic, adipogenic, myogenic and osteogenic molecular markers (Odelberg et al. 2001). The *msx1* gene, however, is not strongly upregulated until the conical blastema stage (Crews et al. 1995; Simon et al. 1995; Koshiba et al. 1998), so whether it is actually involved in initiating dedifferentiation, or maintaining cells in a dedifferentiated state is not clear.

Other genes that might play roles in dedifferentiation are *Nrad*, *radical fringe*, and *notch*. *Nrad* is the ortholog of a diabetes-associated mammalian Ras protein. It is upregulated in the nuclei of newt limb myofibers within 4 h of amputation (Shimizu-Nishikawa et al. 2001). *Rfrng* is expressed in the dorsal ectoderm of amniote limb buds. The position of the AER in these limb buds is specified by the boundary between *rfrng*-expressing dorsal and nonexpressing ventral ectoderm (Laufer et al. 1997; Rodriguez-Esteban et al. 1997). However, in regenerating newt limbs *rfrng* is expressed only in the blastema stem cells and thus appears not to be involved in establishing the position of the AEC (Cadinouche et al. 1999). *Rfrng* activates the *notch* pathway in other developing systems. This pathway is characteristic of undifferentiated and proliferating embryonic cells and adult stem cells (Go et al. 1998), suggesting that in regenerating amphibian limbs, *rfrng*-activated *notch* may be involved in initiating and/or maintaining dedifferentiation.

These genes might regulate the expression of the ECM synthesized by the blastema cells, which in turn could help maintain the dedifferentiated state. Not every ECM molecule known to exist in the limb bud or mature limb tissues has been examined, but enough is known to say that the ECM of the blastema is similar to that of the embryonic limb bud. Fibronectin, tenascin, and hyaluronate are present at high concentrations in embryonic limb buds and are markedly upregulated in the blastema, whereas the expression of laminin is extinguished (Stocum 1995). Laminin is a signature component of basement membranes, which are prevented from re-forming until differentiation of the blastema commences. It is likely that the expression of other components of basement membranes, such as collagen IV and VII, is eliminated as well. Collagen II expression, the trademark of cartilage matrix, is also extinguished, leaving only the expression of collagen I, typical of mesenchymal cells of the limb bud. The expression of sulfated glycosaminoglycans (GAGs) remains the same.

Fibronectin inhibits chondrocyte differentiation and myoblast fusion in vitro, as well as functioning as an adhesive substrate for cell migration (Swalla and Solursh 1984). High concentrations of hyaluronate inhibit chondrogenesis of chick limb bud mesenchyme in vitro, perhaps by interactions with other matrix molecules, or by their ability to bind water, thus expanding the extracellular space and minimizing cell contacts that are essential for differentiation (Toole et al. 1972). Amputated and denervated limbs fail to form a blastema and do not regenerate (see Sect. 4.2). The level of hyaluronate synthesis and accumulation is sharply reduced in these limbs, concomitant with an increased cell density and inhibition of proliferation at the distal tip of the limb. Upon re-innervation of the amputated limb, hyaluronate concentration, cell density and proliferation return to normal levels and regeneration proceeds (Mescher and Cox 1988).

4
Requirements for Proliferation of Blastema Cells

Proliferation of blastema stem cells requires the metabolic effects exerted by endocrine hormones (Liversage 1967; Tassava et al. 1968; Globus and Vethamany-Globus 1985), but is also dependent on specific factors produced by the AEC (Thornton 1968) and the nerves of the blastema (Singer 1952, 1965). Neither the AEC nor innervation by themselves, however, is sufficient for blastema cell mitosis; both are required.

Figure 2 illustrates some of the molecules that are potential mitogens for blastema cells expressed by the AEC and the limb nerves. To have the status of blastema cell mitogen requires that a candidate molecule meet several criteria in addition to being expressed by the AEC or nerves (Brockes 1984). First, it must be secreted from the AEC or nerve endings into the blastema. Second, removal or destruction of the AEC or nerves should result in loss of the molecule from the blastema. Third, the molecule should be able to substitute for the AEC or nerves in maintaining mitosis and/or promoting regeneration to completion. Fourth, selective neutralization of the molecule should abolish its mitogenic effect on blastema cells. Table 1 summarizes how well the candidate molecules expressed in the AEC and the nerves meet these criteria. Most of them meet at least one criterion, but only one, transferrin, so far meets them all. In most cases, the other molecules have been not been tested to see whether they meet the full range of criteria.

4.1
AEC Factors

Preventing contact of freshly amputated urodele limb tissue with the wound epidermis by inserting the limbs into the coelom or folding a

Table 1 Degree to which candidate molecules expressed by the AEC or nerves meet criteria for mitogen status (a question mark indicates that the molecule has not been tested for this criterion)

Tissue	Criterion Secreted into blastema	Removal reduces in blastema	Substitutes for tissue	Neutralization abolishes effect
AEC				
FGF1	?	?	+	+
FGF2	?	?	+	?
FGF8	?	?	?	?
FGF10	?	?	+[a]	?
Nerve				
FGF2	?	?	+	?
GGF2	?	?	+	?
P	?	?	+	+
TF	+	+	+	+

[a] Regeneration-deficient *Xenopus* limbs.
FGF, fibroblast growth factor; GGF, glial growth factor; TF, transferrin; P, substance P.

full-thickness skin flap over the amputation surface prevents blastema formation, even though histolysis and dedifferentiation take place (Goss 1956a; Mescher 1976). DNA synthesis and mitosis of the dedifferentiated cells briefly parallels that of controls until the time when the AEC would normally form, then ceases (Tassava and Garling 1979; Loyd and Tassava 1980). Removal of the AEC at various stages of blastema growth and inserting the limb into the coelom or enclosing the isolated blastema in a fin tunnel leads to truncation of regenerate pattern in the PD axis (Goss 1956b; Stocum and Dearlove 1972), the same effect as observed after removing the AER from chick and frog embryonic limb buds (Saunders 1948; Tschumi 1957). In chick limb buds, truncation has been shown to be due to death of the apical mesenchyme for a distance of 200 μm proximal to the AER (Rowe et al. 1982; Dudley et al. 2002). The effect of AEC removal on blastema cell death and proliferation in vivo has not been investigated, but the ^{3}H-thymidine and mitotic indices of blastemas cultured in the absence of the AEC and in the presence of hormones and nerves, are reduced three- to fourfold (Globus et al. 1980; Smith and Globus 1989). The amputated limbs of regeneration-deficient *Xenopus* also fail to form a fibroblastema when inserted into the coelom (Goss and Holt 1992). Thus the regeneration of even a hypomorphic cartilage spike requires a functional wound epithelium.

In amniotes, FGFs have been shown to be important mediators of survival and proliferation of limb bud mesenchyme (reviewed by Martin 1998). Fibroblast growth factors -2, -4, -8, and -9 are expressed by the AER and FGF-10 is expressed in the mesenchyme. FGF-8 maintains mesenchymal FGF-10 expression and induces the expression of sonic hedgehog (*Shh*) in the posterior mesenchyme. Shh then induces the expression of FGF-4 in the posterior region of the AER. FGF-4 and Shh then maintain each other's expression via a feedback loop. FGF-2 and -4 can substitute for the survival and outgrowth-promoting effect of the AER when applied to the mesenchyme in beads after AER removal (Fallon et al. 1994; Niswander et al. 1993).

FGF-2 and -9 null mice and FGF-4 inactivated mice all express FGF-8 and -10 and form normal limbs, whereas limb buds of FGF-10 null mice fail to form limbs and FGF-8 inactivated mice form shortened limbs with missing digits (Meyers et al. 1998; Lewandowski et al. 2000; Moon and Capecchi 2000; Sun et al. 2000; Moon et al. 2000). These observations suggest that FGF-8 is the most important factor required by the mesenchyme that is secreted by the AER. However, FGF-4 appears to play an important role as well. Inactivation of both the *Fgf-4* and *-8* genes in mouse hindlimb buds by insertion of the *Msx2-cre* transgene results in suppression of limb development (Sun et al. 2002). Limb buds form, but they are very small, due to subnormal recruitment of limb bud cells and proximal cell death. *Fgf-4* and *-8* are not inactivated in the forelimb buds by *Msx2-cre* until after some Fgf-4 and Fgf-8 are produced. These knockout early forelimb buds were normal, but once *Fgf-4* and *-8* were inactivated, proximal cell death took place, reducing their size and resulting in limbs with a normal stylopodium, but a hypoplastic zeugopodium and autopodium. The morphology and gene expression of the AER was normal in both forelimb and hindlimb knockouts until late stages of development, when the AER began to degenerate. Genes normally expressed by the mesenchyme (*Shh, HoxD13, Bmp2* and *4*, and others) could not be detected in hindlimb knockouts, but were normally expressed in forelimb knockouts. There was no detectable effect on cell proliferation. Sun et al. (2002) propose that the AER provides cell recruitment and survival factors that ensure the presence of sufficient mesodermal cells to form a limb.

FGFs also appear to play an important role in the survival and proliferation of the stem cells of the regeneration blastema (Table 1, Fig. 2). The picture of FGF expression is somewhat different in regenerating amphibian limbs than in amniote limb buds. FGF-1 protein is expressed in the mesenchyme and AEC of medium bud stage blastemas (Boilly et al. 1991) and FGF-2 protein has been detected in the AEC and nerves of the mid-bud blastema of axolotls (Mullen et al. 1996). FGF-1 mRNA has also been detected by PCR in medium-bud stage axolotl and newt blastemas, but not by in situ hybridization, suggesting a low level of transcription (Dungan et al. 2002). The receptor for FGF-1 and -2, FGFR1, is expressed in both the AEC and blastema stem cells (Poulin et al. 1993; Poulin and Chiu 1995), and ^{125}I-labeled FGF-1 and -2 bind to sections or membrane preparations of AEC and blastemal mesenchyme (Boilly et al. 1991). Thus it is assumed that FGF-1 or -2 in the AEC is secreted into the subjacent mesenchyme. Such secretion has not been specifically demonstrated, although Chapron (1974) showed that ^{3}H-fucose labeled glycoprotein is synthesized first in the wound epidermis and subsequently appears in the extracellular spaces of the blastema cells. The labeled glycoprotein was not identified, but could be FGFs, fibronectin, which is also synthesized by the basal layer of the AEC (Christensen and Tassava 2000), or other glycoproteins. The expression of FGF-1 and -2 by the *Xenopus* AEC has not been demonstrated. However, the AEC and the blastema stem cells of regeneration-competent *Xenopus* hindlimbs express FGFR1 and -2. Specific inhibitors of the receptors inhibit the regeneration of regeneration-competent limbs in *Xenopus* and the receptors are not expressed by amputated limbs that are regeneration-deficient (D'Jamoos et al. 1998), suggesting that these two FGFs have an important role to play in blastema cell survival and proliferation.

Expression of FGF-4, -8 and -10 has also been demonstrated by PCR in unamputated and regenerating *Ambystoma* limbs (Christensen et al. 2001). All are expressed in unamputated limbs, with FGF-8 having the lowest expression. FGF-4 is downregulated to virtually undetectable levels during regeneration, whereas FGF-8 and -10 expression are upregulated. Levels of FGF-8 expression are low during early stages of dedifferentiation, then increase at early bud to a level that is maintained until late bud. FGF-10 expression increases during dedifferentiation to a level that is maintained until late bud.

The spatial distribution of FGF-8 and -10 in regenerating axolotl limbs has been analyzed by Han et al (2001) and Christensen et al

(2002). FGF-8 and -10 are expressed in both the AEC and the blastema mesenchyme in patterns similar to those of the amniote limb bud, but whether they are secreted from the AEC into the blastema is not known. Expression of FGF-10 is much stronger in the blastema stem cells than in the AEC (Christensen et al. 2002). FGF-8 mRNA is first detectable during dedifferentiation, increases to a peak at medium bud, then disappears by the palette stage (Han et al. 2001). Expression is confined to the basal (functional) layer of the AEC and a thin band of mesenchyme cells just under the AEC; it is strongest in the anterior AEC. Interestingly, while this expression pattern is similar to that of the amniote limb bud, it differs from the expression pattern in the axolotl limb bud, where FGF-8 is expressed primarily in the mesenchyme (Han et al. 2001). In *Xenopus*, FGF-8 is also expressed in the AEC and FGF-10 in the blastema stem cells of regeneration-competent tadpole limbs, but neither factor is expressed in the amputated hindlimbs of regeneration-deficient tadpoles approaching metamorphosis (Christen and Slack 1997; Yokoyama et al. 2000).

FGF-1 and -2 are currently the best AEC candidates to be survival factors and/or mitogens for blastema cells. FGF-1 satisfies two of the four criteria and FGF-2 satisfies one (Table 1). FGF-1 elevates the mitotic index of epidermis-free blastema cells in vitro (Albert et al. 1987). FGF-1 injections or beads promote the accumulation of blastema cells up to a medium bud stage in amputated epidermis-free limbs inserted into the body cavity. These cells exhibit significant cell cycle activity (Dungan et al. 2002). Blastemas release a mitogenic activity into culture medium that is depressed by one-third when the medium is treated with antibody to FGF-1 (Zenjari et al. 1997). FGF-2 has also been found to elevate the mitotic index of regenerating limbs covered with a full-thickness skin flap in vivo (Chew and Cameron 1983).

Exogenous human FGF-8 provides only marginal improvement in the regeneration of regeneration-deficient *Xenopus* limbs (Yokoyama et al. 2001), but this may be because FGF-10 expression is missing in these limbs. Human FGF-10 applied to regeneration-deficient *Xenopus* limbs restores the regeneration of digits while simultaneously inducing expression of FGF-8 in the AEC, but fails to restore more proximal elements. These observations reveal that FGF-8 expression by the AEC in regenerating limbs is probably dependent on FGF-10. However, they do not reveal whether FGF-8 is required for proliferation of blastema cells in regeneration-competent limbs. These experiments need to be repeated with *Xenopus* FGF-8 and -10, and experiments to test whether FGF-8 can substitute for the AEC in promoting blastema cell mitosis in regeneration-competent limbs need to be conducted.

The pathways that regulate the production of FGFs by the AEC are unknown. Several transcription factors associated with limb development are expressed in the AEC (Fig. 2). The genes for the transcription factors Msx2 and Dlx3 are strongly expressed in the early wound epidermis of regenerating axolotl limbs (Mullen et al. 1996; Koshiba et al. 1998; Carlson et al. 1998). Expression of *dlx3* is first detected at the accumulation blastema-early bud stages, peaks at late bud, when it is also detected in the mesenchyme, and decreases to zero by late digit stages of regeneration (Mullen et al. 1996). Three inhibitor of differentiation genes, *Id2, Id3,* and *HES1* are also upregulated in the AEC of regenerating newt limbs (Shimizu-Nishikawa et al. 1999). The role played by these genes in the structure and function of the AEC is not clear.

4.2
Neural Factors

Like epidermal removal, total denervation of an amputated urodele limb by cutting spinal nerves III, IV, and V does not prevent histolysis or dedifferentiation, but a blastema does not form. The regeneration of hypomorphic spikes in regeneration-deficient *Xenopus* limbs is also nerve-dependent (Kurabuchi and Inoue 1983; Filoni and Paglialunga 1990). The effect of the nerves is strictly quantitative; regeneration can be supported by either sensory or motor fibers, as long as the number of fibers is at or above a threshold (Singer 1952).

In amputated and denervated urodele limbs, DNA synthesis by the dedifferentiated cells increases in parallel with controls for a few days, then falls, but the mitotic index never rises above zero, suggesting that the cells cannot complete the cell cycle and divide (Tassava and Mescher 1975). TUNEL staining indicates that dedifferentiated cells in these limbs undergo apoptosis and are removed by macrophages (Mescher et al. 2000). An AEC does not form in simultaneously amputated and denervated limbs, but it is not the lack of an AEC that prevents mitosis, as shown by denervating limbs in which a blastema (and thus an AEC) has already formed. Cell cycle analysis indicates that the blastema cells of these limbs arrest in G_1 (Maden 1979; Oudkhir et al. 1985). The mitotic index falls to zero (Goldhamer and Tassava 1987), but the blastemas are able to undergo differentiation and morphogenesis into miniature regenerates because they have attained a critical mass of cells (Schotte and Butler 1944; Maden 1981; Singer and Craven 1948).

Denervation depresses RNA and protein synthesis without changing the qualitative profile of the proteins synthesized (Dresden 1969; Lebowitz and Singer 1970; Morzlock and Stocum 1972; Singer 1978). Glycosaminoglycan synthesis is strongly reduced (Smith et al. 1975; Mescher and Cox 1988; Young et al. 1989). Hyaluronate synthesis is affected more than any other GAG, consistent with the fact that it is the major GAG synthesized during blastema formation and growth (Mescher and Munaim 1986).

There is some evidence that all blastema cells are not equally nerve-dependent and that the effect of denervation on the majority of the cells might be indirect. Expression of an intermediate filament called 22/18 in regenerating limbs is associated with a subpopulation of blastema cells derived from Schwann cells; this subpopulation is the most sensitive to denervation (Brockes and Kintner 1986). This apparent differential sensitivity has led to the idea that blastema cells originating from Schwann cells might produce an inhibitor of mitosis when their own cycling is halted by denervation, thus halting the proliferation of blastema cells derived from other cell types (Ferretti and Brockes 1991). Support for the existence of such an inhibitor comes from experiments in which pieces of peripheral nerves, implanted into amputated adult newt limbs, delayed regeneration in 80% of the cases (Irvin and Tassava 1998). Implants of other tissues did not have this effect, implying that the nerve implants specifically raised the requirement for proliferation factors by increasing the number of nerve-dependent blastema cells derived from Schwann cells.

What are the nerve mitogenic factors? FGF-1, -8 and -10 appear to be unlikely. Although FGF-1 mRNA is produced by neurons of limb nerves, transcript levels do not increase after amputation, and FGF-1 injections and bead implants do not rescue the regeneration of limbs denervated at early bud (Dungan et al. 2002). Whether FGF-8 and -10 are present in limb nerves is not known. Denervationstrongly downregulates the expression of FGF-8 transcripts in regenerating *Ambystoma* limbs, but only slightly affects the level of FGF-10 mRNA (Christensen et al. 2001). This result suggests that if FGF-8 is involved in maintaining blastema cell proliferation by an autocrine mechanism, its synthesis is controlled by the nerve (Christensen et al. 2001).

Two growth factors and a neurotransmitter present in nerves have been identified that can substitute for the nerves (Table 1, Fig. 2). One of these is the same FGF expressed by the AEC, FGF-2. FGF-2 protein is present in the nerves re-innervating the blastema, is downregulated in the nerves by denervation, and promotes the regeneration of denervated axolotl limbs to digit stages in vivo when delivered to early blastemas in bead implants (Mullen et al. 1996). The second growth factor is glial growth factor 2 (GGF-2), a member of the neuregulin family. Recombinant human GGF-2 infused into denervated nerve-dependent axolotl limb blastemas maintained the ^{3}H-thymidine labeling index at control levels and allowed regeneration to digit stages (Wang et al. 2000), similar to the result observed when spinal ganglia are implanted into nerve-dependent blastemas (Tomlinson and Tassava 1987). A newt neuregulin gene was cloned from spinal cord and was shown to be expressed in newt dorsal root ganglia (Wang et al. 2000). However, the secretion of neither FGF-2 nor GGF-2 into the blastema has been demonstrated, nor have the other two criteria for mitogen status been satisfied by these growth factors.

The neurotransmitter is substance P, a peptide widely distributed throughout the mammalian central and peripheral nervous system that is mitogenic for mammalian connective tissue cells (Nilsson et al. 1985). Substance P is found in the cell bodies of adult newt sensory ganglia and in the AEC and in the mesenchyme of the blastema (Globus and Alles 1990). The peptide is mitogenic for blastema cells in vitro at low concentrations and an antibody to substance P applied to spinal ganglia cultured transfilter to blastemas decreases the mitogenic effect of the ganglia in a dose-dependent fashion (Globus 1988). Whether substance P is secreted into the blastema from the nerves and whether it is eliminated from the blastema by denervation is not known.

The iron-transport protein, transferrin, is the only factor present in limb nerves that meets all the criteria for mitogen status (Mescher 1996). Iron is a cofactor for many enzymes essential for cell proliferation, including the rate-limiting enzyme for DNA replication, ribonucleotide reductase (Mescher and Munaim 1988; Sussman 1989). Transferrin is axonally transported and released distally in the nerves of regenerating axolotl limbs; nerve transection reduces the concentration of blastema transferrin by 50% (Kiffmeyer et al. 1991; Mescher and Kiffmeyer 1992). Transferrin can substitute for the nerve or nerve extracts in maintaining blastema cell proliferation in vitro (Munaim and Mescher 1986;

Albert and Boilly 1988) and in vivo (Mescher and Kiffmeyer 1992). Chelation of ferric ions or immunoabsorption with anti-transferrin antiserum abolishes the mitogenic effect of nerve extracts on blastema stem cells. The inhibitory effect is reversed by adding iron to the chelated extract or purified transferrin to the antiserum-treated extract (Munaim and Mescher 1986; Mescher et al. 1997). The only credential missing from the transferrin portfolio is the demonstration in vivo that it can promote the regeneration of denervated limbs to completion.

4.3 Do the AEC and Nerves Have the Same or Different Functions?

The preceding discussion shows that FGFs are expressed by three sources during regeneration, the AEC, the nerves, and the blastema cells themselves. A further source of growth factors for regeneration would be the ECM, which sequesters growth factors that would be released during histolysis. The FGF-1, -8, and -10 produced by the blastema cells themselves is clearly not sufficient to maintain them in an undifferentiated and dividing state. Additional growth factors produced by the AEC and the nerve together are required. Furthermore, the expression of FGFs and other factors by the AEC and the blastema cells appears to be nerve-dependent. Denervation downregulates the expression of FGF-2 and *dlx3* by the AEC (Mullen et al. 1996) and the expression of FGF-8 by the blastema cells (Christensen et al. 2001). Regeneration of nerve fibers into the blastema is in turn dependent on factors produced by the blastema cells. Regeneration of these fibers is promoted in vitro by factors produced by blastema cells (Richmond and Pollack 1983), some of which are known neurotrophic factors such as brain-derived neurotrophic factor, neurotrophins 3 and 4, glial-derived neurotrophic factor and hepatocyte growth factor/scatter factor, as well as yet unidentified factors (Tonge and Leclere 2000). These unidentified factors could include FGFs.

Other mitogens produced by the AEC and the nerves, not yet investigated, could also be required for blastema cell proliferation. For example, although FGF-2 can act as a competence factor for the fibroblasts of healing mammalian skin wounds, PDGF appears to be the major competence factor involved in the acquisition of competence, and it must interact with two progression factors, IGF-I and EGF, to drive fibroblasts to the G_1/S transition (Morgan and Pledger 1992). Are these growth factors also involved in the cycling of blastema cells?

What are the specific roles of the AEC and the nerve in maintaining blastema cells in an undifferentiated and dividing state? Are they the same or are they different? The simplest hypothesis is that they are the same: blastema cells have a quantitative threshold requirement for mitogens from the start of regeneration that is met through production of the mitogens by both the AEC and the nerve. A linear scheme in which the nerve induces the synthesis of mitogens by the blastema cells, which then act on the AEC to stimulate the expression of additional mitogens is unlikely, because the nerves and the wound epidermis both appear to be functional (i.e., are required) from the beginning of regeneration (Stocum 1995) and the mitogens hypothetically would be available very shortly after amputation as a result of ECM degradation. Currently, FGF-2 is the only mitogen known to be expressed by both the AEC and the nerves. The precise cellular role of FGF-2 in limb regeneration is unknown. It might act as a competence factor. Alternatively, FGF-2 might play a role in keeping blastema cells in an undifferentiated state by regulating the expression of transcription factors such as msx-1 or the expression of ECM molecules such as hyaluronate, fibronectin, and tenascin. Or, FGF-2 might play a role in both the acquisition of competence and maintenance of the dedifferentiated state.

The same-function hypothesis could explain the fact that dependence on the nerve for regeneration arises during the digital stages of limb development, when the limb bud is becoming heavily innervated. Acquisition of nerve-dependence in urodeles is associated with expression of the 22/18 antigen in the blastema (Fekete and Brockes 1987, 1998). By contrast, the limbs of aneurogenic embryos, made by removing the neural tube, never become dependent on the nerve for regeneration (Yntema 1959a, 1959b) and their regeneration blastemas do not express 22/18, though they remain dependent on the AEC (Fekete and Brockes 1988). The level of FGF-10 mRNA expression in the medium bud blastemas of regenerating aneurogenic limbs remains the same as in the unamputated limb, but the expression of FGF-8 transcripts is upregulated, as in neurogenic limbs (Christensen et al. 2001). Aneurogenic limbs allowed to become innervated acquire nerve dependence and exhibit a 22/18-reactive subpopulation of cells in their regeneration blastemas (Fekete and Brockes 1987). Interestingly, when reciprocal combinations of skin and internal tissues were made from neurogenic and aneurogenic limbs, the combination of neurogenic skin and aneurogenic internal tissues failed to regenerate, whereas the combination of aneurogenic skin and neurogenic internal tissues regenerated normally (Steen and Thornton 1963).

These results suggest that during limb development, either the requirement of blastema cells for the same mitogens increases and is satisfied by additional mitogen production by the nerves, or remains the same but the AEC expresses the mitogens at lower levels, requiring the nerves to make up the difference. It would be instructive to compare the levels of FGF-2 in the AEC of neurogenic and aneurogenic regeneration blastemas. If the requirement of blastema cells for FGF-2 increased as the limb buds differentiated, the AEC of the aneurogenic limb presumably would have to increase FGF-2 production, whereas FGF-2 production by the AEC of a neurogenic limb would stay the same. If the requirement of blastema cells for FGF-2 remained the same, but the AEC expressed FGF-2 at lower levels, the level of FGF-2 production by the aneurogenic AEC would not change, but it would decrease in the AEC of the neurogenic limb.

There is some evidence from studies on regenerating *Xenopus* hindlimb buds, that there is a loss of mitogen production by blastema cells that is made up by the nerves. Even as regenerative capacity declines, the developing hindlimb buds of *Xenopus* acquire dependence on the ingrowing nerves for whatever regenerative capacity remains (Filoni et al. 1991; Cannata et al. 1992). Implantation of regeneration-competent (nerve-independent) limb bud tissue into denervated, regeneration-deficient (nerve-dependent) limbs maintained blastema cell proliferation at levels comparable to limbs implanted with spinal ganglia, whereas implants of late (nerve-dependent) limb bud tissue did not, suggesting that the cells of the nerve-independent limb bud produce mitogenic factors, but stop producing these factors as they differentiate, becoming dependent on the nerve for them (Filoni et al. 1999). Like the loss of regenerative capacity, the change from nerve independence to dependence is due to intrinsic changes in the limb cells themselves, not to an influence of the ingrowing nerves. Early hindlimb buds grafted to denervated late hindlimb buds and then amputated as they reach progressively later stages of development nevertheless become nerve dependent as they differentiate in the absence of nerves (Filoni et al. 1991).

Alternatively, the requirements of the blastema cells for mitogens might change qualitatively and the nerves might supply mitogens that the AEC cannot, such as transferrin, substance P, and GGF-2, and others. The cells of epidermis-free newt limb blastemas cultured in vitro transfilter to spinal ganglia arrest in G_0 and differentiate prematurely into cartilage (Globus et al. 1980), whereas the cells of denervated blastemas with AEC intact appear to arrest in G_1, S or possibly G_2 (Maden 1979; Tassava and Mescher 1975), suggesting potentially different functions of the AEC and nerves.

Clearly, we are far from understanding which molecular functions of the nerve and the AEC are collaborative and which might be tissue specific.

5 Developmental Plasticity of Dedifferentiated Cells

5.1 Plasticity of Adult Mammalian Reserve Stem Cells

Recent experiments have shown that the reserve stem cells of mammals have a developmental potential (prospective potency) much greater than their normal fate (prospective significance), giving rise to the exciting possibility that a single stem cell type could be used to regenerate a wide range of tissues. This potency has been revealed by in vitro experiments in which enriched stem cell populations, or clonally-derived stem cells, are co-cultured with other cell types, and by in vivo experiments where marked stem cells are injected into X-irradiated and severe combined immunodeficient (*scid*) mice, or into blastocysts to form chimeras (Fuchs and Segre 2000; Weissman 2000b, Blau et al. 2001; Stocum 2001). Thus, clonally derived neural stem cells differentiated into blood, hepatocytes, intestinal cells, skeletal muscle, and cardiac muscle when injected into embryos (Clarke et al. 2000). Highly enriched hematopoietic stem cells injected into dystrophic mice contributed to regenerated muscle (Gussoni et al. 1999) and differentiated to hepatocytes in regenerating liver (Lagasse et al. 2000). A single bone marrow cell, derived by limiting dilution, was shown to home to bone marrow in lethally irradiated mice and repopulate the hematopoietic system (normal fate), but also to differentiate into epithelial cells of the liver (preductule cells), digestive tract, lung, kidney tubules, and skin (Krause et al. 2001).

5.2 Plasticity of Blastema Cells

The plasticity of limb regeneration blastema cells has been tested by tissue grafting experiments. When triploid limb skin or cartilage was grafted in place of the limb skin and cartilage of diploid axolotls, triploid dermal fibroblasts and chondrocytes dedifferentiated to become part of the blastema. The blastema cells derived from dermal fibroblasts differentiated into both fibroblasts and chondrocytes; blastema cells derived from chondrocytes differentiated into chondrocytes and fibroblasts of the connective tissues of the perichondrium, joints, dermis, and skeletal muscle (Namenwirth 1974; Dunis and Namenwirth 1977; Holder 1989; Mescher 1996). Blastema cells derived from cartilage implanted into X-irradiated amputated limbs have been reported to differentiate into muscle cells, as well as into the skeletal elements and connective tissues (Maden and Wallace 1975; Desselle and Gontcharoff 1978). Blastema cells derived from marked muscle cells have been found to differentiate into chondrocytes and other cell types after being implanted into amputated wild-type *Xenopus* or newt limbs (Steen 1973; Casimir et al. 1988; Lo et al. 1993; Kumar et al. 2000). In summary, fibroblasts and myogenic cells appear able to become chondrocytes, and chondrocytes can become fibroblasts and myogenic cells. Myogenic cells have not been observed to become fibroblasts or vice versa, though fibroblasts could hypothetically become muscle cells by first becoming chondrocytes during one round of regeneration, and could then become muscle cells after a second round.

We do not know whether blastema stem cells, or a subpopulation of these cells, have individual developmental potency beyond the range of cell types present in the limb. This is an interesting question that can be related to the origin of stem cells in the marrow of endochondral bones in the regenerating limb. The development of embryonic endochondral limb bones is characterized by a well-defined series of steps leading to ossification (Olsen 1999; Ducy 2000; Ducy et al. 2000; Vortkamp 2001). The mesenchyme cells of the limb bud condense to form a cartilage template surrounded by a periosteum, which forms a thin shell of periosteal bone around the cartilage. The chondrocytes proliferate, hypertrophy, and undergo apoptosis within a calcified matrix. Osteoclasts hollow out a marrow cavity within the calcified matrix that becomes populated by hematopoietic and mesenchymal stem cells (HSC and MSC, respectively). HSCs and MSCs enter the cartilage template as it is invaded by blood vessels. The MSCs originate in the periosteum, and the HSCs in extraskeletal sites, such as the liver (Gilbert 2000). Some of the MSCs adhere to the cartilage matrix and differentiate into osteoblasts, while others, along with the HSCs, take up residence in the marrow cavity. Fractured long bones are healed in the same series of steps. Mesenchymal stem cells in the periosteum and bone marrow proliferate in the break to form a soft callus. The callus cells condense to form a cartilage template bridging the fracture site, which goes through the same sequence of differentiations as an embryonic endochondral template (Einhorn 1998).

During adult urodele limb regeneration, the endochondral skeletal elements presumably differentiate from the blastema in the same way as endochondral bones in the limb bud or a healing endochondral fracture. Assuming this is so, what is the origin of HSCs and MSCs in the marrow cavity of regenerating limbs? Do they accompany blood vessels as they invade the calcified cartilage template? If so, some of the blastema cells, particularly those that become MSCs in the periosteum, may have a developmental potency extending beyond limb cell types that could be tested by dissociating them and grafting them into early embryos. Or, are MSCs and HSCs derived by the division of existing stem cells in the marrow cavity and periosteum of adjacent differentiated tissues, which then migrate into the marrow cavities of the regenerated long bones?

5.3
Do Reserve Stem Cells Contribute to the Blastema?

The extent of blastema cell plasticity gives rise to the question of whether reserve stem cells contribute to the blastema. Might MSCs and HSCs residing in the marrow and periosteum repopulate regenerating long bones by first becoming part of the blastema and then becoming new stem cells or differentiating into other cell types of the limb? Since limb muscle also contains reserve stem cells ('satellite cells') that repair injured muscle in the absence of amputation, the same question may be asked of these cells. A significant mechanistic difference between tissue regeneration (fractured bone or torn muscle) and the regeneration of the limb as a whole would be indicated if reserve stem cells were used only in the process of individual tissue regeneration, while whole limb regeneration depended primarily on stem cells created by dedifferentiation.

A contribution of reserve stem cells to the blastema has never been ruled out, but there is no direct evidence for such a contribution. Testing for it is difficult because a way is required to selectively label mesenchymal stem or satellite cells so they can be tracked during limb regeneration. Muscle explants might possibly be used to label satellite cells. The nuclei of satellite cells in newt muscle explants are selectively labeled by ^{3}H-thymidine during the first 6 days of culture (Cameron et al. 1986). Such labeled muscle explants might be transplanted back into an amputated limb and the labeled cells subsequently followed. One problem with interpreting the results of this experiment, however, might be that the satellite cells are activated by explanting the muscle, whereas we do not know whether such activation takes place after amputation of the limb in vivo. Salamanders with marked bone marrow cells might be made by repopulating the marrow of irradiated animals with marked marrow cells (triploid, transgenic for LacZ or green fluorescent protein, opposite sex chromosomes, lipophilic tracker dyes), or by grafting marked hematopoietic regions of early embryos to unmarked hosts. The limbs of the marked animals could then be amputated and the blastema examined for the presence of marked cells.

6
Spatial Organization of Tissue Patterns in the Regenerating Limb

6.1
The Blastema is a Self-Organizing System

Although individual blastema stem cells have histogenetic plasticity, the blastema as a whole is determined from its inception. Grafting experiments on *Ambystoma* larvae show that the blastema self-organizes its tissue patterns, as opposed to being patterned by a set of inductive stimuli from the mature tissues adjacent to the blastema (Stocum 1968a, 1968b, 1984, 1996, 2000). Blastemas grafted to the dorsal fin, exchanged between forelimb and hindlimb, grafted to a different PD level, or rotated around the PD axis to disharmonize the AP and/or DV axes of blastema and adjacent tissues, always develop according to origin with regard to limb type, PD level, and handedness. The fibroblasts and muscle of the limb carry the information for the self-organizing pattern (Carlson 1974, 1975; Namenwirth 1974; Holder 1989). The central importance of fibroblasts to patterning is most clearly shown by experiments in which

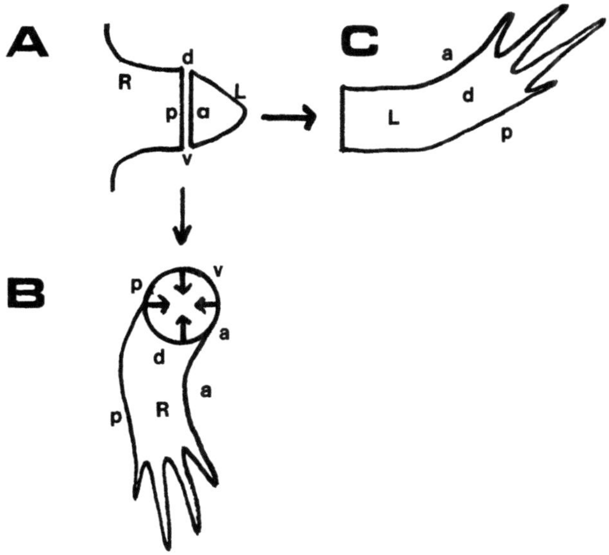

Fig. 3 **A** Supernumerary limb formed by grafting a blastema from a left limb (*L*) to a right limb (*R*) with reversal of the anteroposterior axis. **B** This operation creates a discontinuity between anterior and posterior positional identities on both sides of the construct, resulting in two extra complete circumferences (*d*, dorsal; *a*, anterior; *v*, ventral; *p*, posterior), of which only one is shown. Intercalation (*short arrows*) takes place along the radii of the circumference and the cells distalize to create a right-handed supernumerary regenerate. **C** The grafted blastema develops as a left-handed regenerate

unirradiated skin was grafted to irradiated limbs in axolotls. The irradiated host muscle and cartilage tissues cannot contribute to the regeneration blastema that forms after amputation of these limbs. Nevertheless, the regenerate that forms, though lacking in muscle, has a normal skeletal pattern, indicating that the blastema cells derived from the dermal fibroblasts of the unirradiated skin were capable of organizing this pattern (Namenwirth 1974).

When a distally derived blastema is grafted to a more proximal level, the cells at the proximal level dedifferentiate and distalize to fill in the missing intermediate structures, a phenomenon called intercalary regeneration (Stocum 1975; Iten and Bryant 1975; Pescitelli and Stocum 1980). Likewise, when the AP or DV axis of the blastema is reversed with respect to the adjacent tissues, confronting anterior and posterior, or dorsal and ventral half-circumferences of cells, supernumerary limbs are generated at the sites of confrontation by intercalary regeneration within the complete limb circumferences thus created, followed by proximodistal outgrowth (Carlson 1974, 1975; Bryant and Iten 1976; Cameron and Fallon 1977; Tank 1978; Holder and Tank 1979; Stocum 1980a; Fig. 3). Both host and graft tissue contribute to the supernumerary limb to varying degrees (Stocum 1982), indicating that cells of the half-circumference contributed by the limb stump undergo dedifferentiation, just like the proximal cells in a distal blastema to proximal stump graft.

Self-organization is a property of blastema cells from their inception, as shown by experiments in which proximal halves of forelimb blastemas, derived by amputation through the mid-humerus and just starting to undergo differentiation, were grafted to the dorsal fin (Stocum 1968b), or to the tarsal level of the hindlimb (Stocum and Melton 1977). The grafts are re-covered with wound epidermis, forcing its cells to dedifferentiate again. The dedifferentiated cells nevertheless self-organize perfectly into forelimb regenerates. In the experiment in which the graft was placed at the level of the tarsus, no intercalary regeneration took place, despite the large discontinuity between graft and host limb levels. There is thus a preferred distal-to-proximal polarity to the recognition of a structural discontinuity and intercalary regeneration in the PD axis. The basis for this preference is unknown.

6.2 A Mechanism of Self-Organization Based on Inherited Positional Boundaries and Intercalary Regeneration

The development of the blastema according to origin, even after its cells are forced to dedifferentiate afresh, suggests a model of self-organization in which blastema cells inherit from their parent limb cells a memory of their position in the limb (Fig. 4). The limb can be viewed as a field of cells, each of which has a 'positional identity' in three-dimensional space. These positional identities constitute a 'normal neighbor' map (Mittenthal 1981). During regeneration, the proximal base, distal tip, and circumference of the map represents a set of boundary positional identities that are used as reference points to restore a full map within the boundaries by intercalary regeneration. The progeny of the initial blastema cells can change their positional identities only in a distal direction along the PD axis (rule of distal transformation) and in any direction along the radii within the circumferential boundaries. Interac-

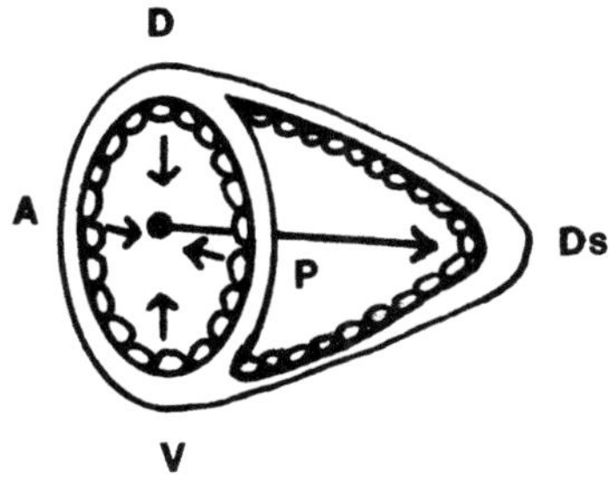

Fig. 4 Model for restoring map of positional identities lost by amputation. The shell of blastema cells under the wound epidermis carries circumferential boundary positional identities. Those cells at the base of the blastema carry proximal boundary identities and the cells in contact with the AEC carry distal boundary identities. Interclation can take place within any gaps in the circumference, along the radii of the circumference (*short arrows*) and between proximal and distal boundaries (*long arrow*). *A*, anterior; *P*, posterior; *D*, dorsal; *V*, ventral; *Ds*, distal

tions between cells with non-neighbor positional identities lead to the intercalation of missing positional identities until a complete map is reconstituted (French et al. 1976; Maden 1977; Mittenthal 1981; Stocum 1978a, 1980b, 1982, 1996). As the map is restored, the system becomes stable and differentiation in the spatial pattern specified by the map takes place.

The interactions between any two cells with non-neighboring positional identities follow the 'rule of least (shortest) intercalation', which states that given a choice of intercalating the greater or lesser number of positional identities that will restore a complete map, cells will always choose the lesser number, or shorter route (French et al. 1976). The rule of least intercalation appears to be a biological analogue of the principles of least action and least time in physics. The principle of least action states that an object going from one point to another takes the path over which the average kinetic energy minus the potential energy is least (Feynman et al. 1977a). The principle of least time (Fermat's principle) states that, out of all the possible paths light might take to go from point A to point B, the path taken is the one which requires the shortest time (Feynman et al. 1977b). These 'least principles' appear to a feature of physical systems on many, if not all, levels of organization.

The experimental evidence shows that intercalary regeneration takes place to fill in gaps bounded by distal and proximal blastema stem cells after a distal to proximal shift-level graft. The base of the grafted distal blastema provides the distal boundary for intercalary regeneration in these experiments. Assuming that intercalation is the mechanism by which PD positional identities are restored during normal regeneration, how is the distal boundary of a self-organizing blastema established following simple amputation? The AEC might be involved in this process. The AEC does not appear to possess any asymmetric positional information in its circumference, since rotating the epidermis with respect to underlying tissues does not result in any pattern abnormalities during regeneration (Carlson 1974, 1975), but it does confer a distal direction of PD outgrowth on the blastema stem cells. Shifting the AEC to an eccentric location by removing a patch of skin at the base of the blastema in *Ambystoma* larvae results in outgrowth of the regenerate at an angle to the stump; grafting an extra AEC to the base of the blastema results in production of a supernumerary limb (Thornton 1960, 1962; Thornton and Thornton 1965).

The difference in pattern-forming ability between established medium bud-stage blastemas of newts or *Ambystoma* larvae that are denervated in situ vs. grown in fin tunnels in the absence of epidermal contact also suggests that the AEC might be involved in creating the distal boundary of the regenerate. Growth is halted in both cases and the blastemas form miniature skeletal elements. However, the skeletal pattern is complete in the PD axis in the regenerates denervated in situ, (Singer and Craven 1948; Schotte and Butler 1944; Maden 1981), whereas it is distally truncated at the level of the carpals in the implanted epidermis-free blastemas (Stocum and Dearlove 1972). This difference suggests that the AEC is essential to PD patterning in a way that is distinct from its growth-promoting function.

The inheritance of a proximal boundary by blastema cells and the possible role of the AEC in establishing and maintaining a distal boundary, suggests a mechanism of blastemal self-organization. Those cells in contact with the wound epidermis during initial dedifferentiation will assume a distal boundary function. It is also possible that following amputation, blastema cells with either proximal or distal boundary identities are created at random during initial dedifferentiation and that the distal cells sort out from the proximal cells to associate with the AEC because they and the AEC represent normal neighbors. Alternatively, the genes for distal and proximal might both be expressed in every dedifferentiated cell, but proximal genes are shut off in those cells in contact with the AEC (see Sect. 6.3.1). Regardless, the starting point for self-organization would be the creation of proximal and distal boundary cells that are juxtaposed during initial dedifferentiation. Intercalation of intermediate positional identities would then take place between these distal and proximal blastema cells as they proliferate.

This mechanism allows for the very early representation of all the amputated segments of the limb. For example, if the forelimb were amputated at the level of the body wall, removing the humerus, radius and ulna, and hand, positional identities for each of these segments could be intercalated very rapidly between the proximal and distal boundaries, even prior to the accumulation blastema stage. Intercalation of intrasegmental positional identities would follow and, along with proliferation, would expand each of the segments. The initial size and rate of expansion of each segment are unknown. Truncation of the PD pattern at the level of the carpals in epidermis-free medium bud blastemas implanted into fin tunnels could be due to loss of maintenance of distal boundary functions of apical cells or loss of proliferative capacity to expand a small autopodial segment. It could also be due to apical cell death, even though the blastemas are supposedly in an environment supportive of cell survival. Cell death and proliferation studies on epidermis-free blastema implants would be helpful in determining the mechanism of pattern truncation.

A similar mechanism has been proposed for the self-organization of the embryonic urodele limb disc (Stocum and Fallon 1982) and may be operating in the limb discs of amniotes as well. Stark and Searls (1973) showed that all three limb segments of the chick wing bud appear to be present in the bud just after its appearance, when it is a low mound of cells. They mapped the prospective significance of ^{3}H-thymidine labeled blocks of donor mesenchyme cells transplanted into stage 18–24 wing buds and found that all of the limb segments were represented at stage 18–19. This finding was confirmed recently by Dudley et al (2002). They injected DiI into small groups of stage 19 wing bud mesenchymal cells just under the AER and at different distances proximal to the AER. Cells labeled just under the AER formed the autopodium and cells labeled 100 and 200 μm proximal to the AER formed the zeugopodium and stylopodium, respectively. The apex of the bud was able to self-organize into digits when transplanted ectopically, showing that the apical mesenchyme is destined to become digits. Dissociation of the apical mesenchyme and repacking it into an ectodermal jacket, however, showed that it is not determined as digits until stage 24. Apical cells from stage 20 formed all three segments, whereas apical cells from stage 22 formed zeugopodium and autopodium, and cells from stage 24 formed only autopodium. These observations suggest that determination of segments takes place in proximal-to-distal sequence. Similar mapping experiments need to be conducted on regeneration blastemas.

The polar coordinate model of intercalation (French et al. 1976) proposed that distal transformation required a complete circumference of positional identities (complete circle rule). Consistent with this idea, proximodistal outgrowth of the blastema does not take place if the positional identities of cells around the periphery of the limb are identical. This was shown by an experiment in which irradiated newt limbs received grafts of longitudinal strips of unirradiated skin containing a single circumferential positional identity (say, dorsal). The skin strips were rotated 90° and wrapped around the circumference of the limb. Limbs amputated through the graft failed to form a blastema and healed by scarring whereas irradiated limbs that received grafts around their circumference of un-irradiated skin taken from all four quadrants of the limb regenerated (Lheureux 1975). However, unirradiated skin from two opposite quadrants also allowed regeneration, contravening the complete circle rule. Experiments with surgically constructed double half anterior and posterior hindlimbs of *Ambystoma* larvae also showed that a complete circumference is not required for distalization. Such limbs can distalize after simple amputation or by intercalary regeneration after grafting a normal wrist blastema to a double half limb stump made at the level of the humerus or vice versa (Stocum 1978b, 1981, 1983; Slack and Savage 1978).

6.3 The Use of Retinoids to Investigate the Physical Basis of Positional Identity

Retinoic acid (the acid derivative of vitamin A) administered systemically during the stage of initial blastema formation proximalizes, posteriorizes, and ventralizes the positional identities of blastema cells in a concentration-dependent manner (Niazi and Saxena 1978; Maden 1982, 1997; Thoms and Stocum 1984; Niazi et al. 1985; Kim and Stocum 1986a; Ludolph et al. 1990, 1993a; Stocum and Mitashov 1990; Stocum and Maden 1990; Stocum 1991; Monkmeyer et al. 1992; Niazi 1996). That is, cells take on positional identities more proximal, posterior and ventral to the identities of their position of origin. At the maximum concentration of RA, the cells of a wrist-level blastema will be proximalized to the level of the shoulder and a whole limb, complete with shoulder girdle, will regenerate from the level of the wrist. In frogs and toads, posteriorization and ventralization results in the development of multiple limbs (Niazi and Saxena 1978; Maden 1983; Niazi 1996). Multiple limbs are not formed in urodeles, but posteriorization and ventralization can be demonstrated in RA-treated anterior and dorsal half or double half limbs (Kim and Stocum 1986b; Stocum and Crawford 1987; Ludolph et al. 1990). The blastema of an amputated anterior or dorsal half limb is not only proximalized, but completes itself in the AP or DV axis. Double half anterior and ventral limbs form two AP-complete mirror-image limbs, each of which is also proximalized. The positional respecification effect is on the blastemal stem cells, not the wound epidermis. Reciprocal exchange of RA-treated vs. untreated blastema mesenchyme and epidermis in larval axolotls showed that proximalization occurred only in the combination of treated mesenchyme/untreated epidermis (Maden 1984).

The ability of RA to alter positional identity has been used to probe the physical nature of positional identity. In vitro and in vivo assays indicate that the adhesivity of blastema cells is graded along the PD axis. Proximal blastema cells surround distal ones when the blastema cells are pressed together at their bases and cultured in hanging drops (Nardi and Stocum 1983). Consistent with Steinberg's (1978) differential adhesion hypothesis, this result suggests that blastemas derived from distal levels are more adhesive than those derived from proximal levels. The differential affinity of blastema cells derived from different PD levels was also demonstrated by an 'affinophoresis' assay in vivo (Fig. 5). In this assay, axolotl blastemas derived from the wrist or elbow levels of the forelimb were grafted to the blastema–stump junction of a hindlimb host regenerating from the mid-thigh (Crawford and Stocum 1988a; Egar 1993). The grafted blastemas were displaced distally (i.e., sorted out) to their corresponding level on the host regenerate (ankle and knee, respectively), where they differentiated. Since all interactions between cells take place at their surfaces, the graded affinity of blastema cells for one another suggests that positional identity is encoded in the cell surface.

A direct correlation between positional identity and blastema cell affinity was demonstrated by using RA in combination with the affinophoresis assay or an intercalation assay (Crawford and Stocum

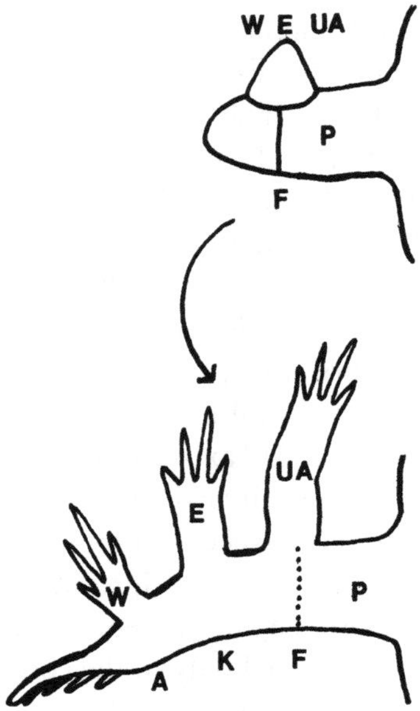

Fig. 5 Affinophoresis of wrist (*W*), elbow (*E*), or mid-upper arm forelimb blastemas (*UA*) grafted to the blastema-stump junction of a hindlimb regenerating from the level of the mid-femur (*F*). View from the posterior (*P*) side of the hindlimb. The grafted forelimb blastemas locate to the level of the hindlimb regenerate that matches their level of origin on the forelimb and differentiate. Thus the wrist regenerate locates to the ankle (*A*) level of the host, the elbow regenerate to the knee (*K*), and the upper arm regenerate remains at the level of the mid-femur. Treatment of wrist or elbow blastemas with RA prior to grafting abolishes distal movement. The grafts are proximalized and remain at the mid-femur level

1988a, 1988b). When the affinophoresis assay was performed using donor wrist and elbow blastemas proximalized by RA, distal displacement was abolished (Crawford and Stocum 1988a). Similarly, grafting an RA-treated wrist-derived blastema to a more proximal level abolished the intercalary regeneration that would normally have taken place between distal and proximal levels (Crawford and Stocum 1988b). These results argue that positional memory is a property of the cell surface and that RA alters gene activity affecting the qualitative and/or quantitative molecular composition of the cell surface. Consistent with this idea is the fact that tunicamycin, which inhibits biosynthesis of the oligosaccharide component of N-linked glycoproteins on the cell surface, blocks the proximalization of positional identity by RA (Johnson and Scadding 1992).

Experiments with chick limb buds suggest that positional identity of limb bud cells is also encoded in the cell surface and that several types of surface receptors and ligands are the basis for identity. Mesenchyme cells from the progress zone of chick limb buds of different stages, or from proximal and distal regions of the same stage, sort out from one another in vitro or in recombinant limb buds in vivo (Wada et al. 1993; Wada and Ide 1994; Ide et al. 1994, 1998). Ephrin A receptors and ligands are expressed in limb buds and an antibody to the EphA4 receptor inhibits sorting, suggesting that the ephrin A family of ligands and this receptor play a role in effecting PD positional identity (Wada et al. 1998). The ephrin A ligands are anchored to the cell surface by glycosylphosphatidylinositol; treatment of chick limb buds with bacterial phosphatidylinositol-specific phospholipase C prevents sorting. Sorting is also inhibited by an antibody to N-cadherin, which accumulates in the distal mesenchyme of the limb bud as development proceeds (Yajima et al. 1999). Finally, RA appears to render distal limb bud cells capable of forming more proximal structures (Tamura et al. 1997). A bead containing RA induces the AP duplication of digits when implanted in the anterior margin of prospective autopodial mesenchyme of the limb bud at stage 20. RA-treated stage 24 distal cells (prospective autopodium) did not sort out from stage 20 distal cells (prospective zeugopodium and autopodium). Furthermore, when a small graft of RA-treated stage 20 mesenchyme was implanted into the apex of a stage 19 mesenchyme, the cells were found in the zeugopodium and autopodium, suggesting that some of them had undergone affinophoresis and migrated proximally.

6.3.1 Nuclear Receptors and Cytoplasmic Binding Proteins Mediate the Biological Effects of Retinoids

The biological effects of retinoids are mediated by nuclear receptors of the steroid/thyroid/vitamin D3 superfamily of transcription factors, the retinoic acid receptors (RARs) and the retinoid X receptors (RXRs) (Manglesdorf et al. 1994). All-trans RA is the specific ligand for the RARs and 9-cis-RA is the specific ligand for the RXRs. When activated by RA, these receptors bind to retinoic acid response elements (RAREs) in the regulatory regions of target genes. Three major RAR isoforms, α,

β, and δ, have been identified in the regenerating newt limb (Maden 1997; Hill et al. 1993; Ragsdale et al. 1989, 1992a, 1992b). The α and β isoforms are expressed uniformly throughout the blastema (Giguere et al. 1989). Delta has two isoforms, of which $\delta 1$ is the most strongly expressed in the blastema (Ragsdale et al. 1992b).

An anterior-to-posterior gradient of all-trans RA has been detected in axolotl blastemas, and proximal blastemas of newts have been found to have higher RA activity than distal blastemas (Brockes 1992; Scadding and Maden 1994). Furthermore, the wound epidermis of distal blastemas has a higher content of 9-cis-RA than the wound epidermis of proximal blastemas, suggesting a role for RXRs in regeneration (Viviano et al. 1995). The presence of RA and its receptors in the blastema, the asymmetric distribution of RA in the blastema, and the ability of RA to alter positional identity, argue strongly for a role of endogenous RA in patterning of the axolotl blastema. This role has yet to be specifically defined, however.

Access of RA to the nuclear receptors may be regulated by cytoplasmic retinoic acid binding proteins (CRABPs) I and II. CRABPs are expressed at high levels during early stages of regeneration in axolotl limbs (Keeble and Maden 1986; McCormick et al. 1988). CRABP I transcripts have been detected in combined muscle and skin of unamputated axolotl limbs by Northern analysis, but not in the regeneration blastema (Ludolph et al. 1993b), suggesting that CRABP expression might be down-regulated in the blastema, freeing more RA for interaction with nuclear receptors.

6.3.2
The $\delta 2$ RAR Mediates Proximalization of Positional Identity

The $\delta 2$ RAR has been identified as the receptor that mediates RA-induced proximalization of positional identity. The identification was made by a transfection experiment using chimeric receptor constructs, consisting of the DNA binding regions of the individual RARs coupled to the ligand binding region of the thyroid hormone (T_3) receptor (Pecorino et al. 1996; Fig. 6). This allows the experimenter to selectively activate the receptor by administering T_3. The chimeric receptor and an alkaline phosphatase reporter gene coupled to a RARE were biolistically transfected into a hindlimb blastema derived by amputation through the ankle. The transfected blastema was then grafted to the mid-humeral

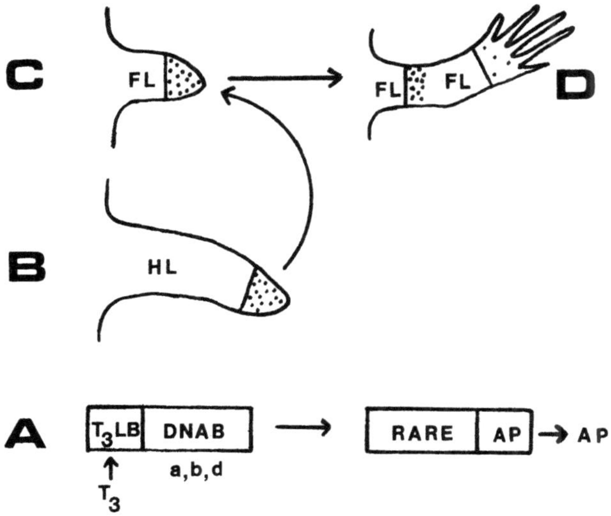

Fig. 6A–D Experiment to determine which retinoic acid receptor mediates proximalization of positional identity of blastema cells. **A** Chimeric receptors were constructed by linking the ligand binding region of the T_3 receptor (*T_3LB*) to the DNA binding regions (*DNAB*) of the α, β, and δ receptor isoforms. A reporter gene was constructed by coupling a retinoic acid receptor element (*RARE*) to the alkaline phosphatase gene. **B** Each chimeric receptor and the reporter were biolistically co-transfected (1/100 cells) into a hindlimb (*HL*) blastema derived from the level of the ankle (*stippled*). **C** The hindlimb blastema was grafted to a more proximal level of the forelimb (FL). **D** The chimeric receptor was activated by injecting the animals with T_3, allowing the receptor to bind by its DNAB region to the RARE of the reporter. Intercalary regeneration took place, resulting in a regenerate composed of a foot and intermediate structures derived from the host forelimb. When the $\delta 2$ DNAB region was used to make the chimeric receptor, some cells of the graft (*stippled*) were found toward the base of the intermediate structures, indicating that this receptor mediates RA-induced proximalization

level of the forelimb and the animals treated with T_3 to activate the chimeric receptor, which in turn activates the reporter.

As expected, grafting a nontransfected ankle blastema to a more proximal forelimb level resulted in a regenerate consisting of a foot derived from the graft and intermediate structures derived by intercalary regeneration from the host forelimb. Transfected blastemas gave the same result as controls when the α, β, and $\delta 1$ DNA binding regions were used to make the chimeric receptor; i.e., the cells expressing the reporter were confined to the regenerated foot and the rest of the regenerate was derived from the host forelimb, indicating that these RARs did not mediate proximalization. By contrast, when the DNA binding region of the $\delta 2$ RAR was used to construct the chimeric receptor, many alkaline phosphatase-expressing cells were found toward the base of the intercalated intermediate structures, indicating that their adhesivity (positional identity) had been proximalized by T_3 activation. It is important to point out that a completely proximalized ankle blastema (i.e., one in which every cell had been transfected) would have elicited no PD intercalation (Crawford and Stocum 1988b). However, the transfection efficiency in this experiment was 1 per 100 cells. Thus 99% of the graft cells were still distal and evoked intercalary regeneration. The 1% that were proximalized behaved as in an affinophoresis assay, migrating proximally to a position more compatible with their new positional identity. The result of this experiment also shows that blastema cells are individually proximalized by RA.

6.4
Transcription Factors and Signaling Molecules Involved in Specifying Positional Identity

In amniote limb buds, distal, posterior, and dorsal signaling centers appear to provide the molecular signals that direct the outgrowth and three-dimensional patterning of the bud. The first of these is the AER, which controls growth and patterning along the PD axis. The second is the zone of polarizing activity (ZPA), located on the posterior margin of the limb bud, that directs patterning in the AP axis (Fallon and Crosby 1977; Johnson and Tabin 1997). The third is the dorsal ectoderm, which directs patterning along the DV axis (Akita 1996). The signals provided by these centers activate transcription factors in cells that in turn specify their positional identity. We have only a rudimentary understanding of how tissue patterns are generated, but we know some of the signals and transcription factors involved and we know that the signaling centers interact in an interdependent way to generate the pattern. A current model for amniote limb buds (Martin 1998; Vogt and Duboule 1999) is that FGF-8 induces the expression of FGF-10 in the prospective limb cells of the lateral plate mesoderm, which in turn induces FGF8 expression in the overlying ectoderm to establish the AER. The position of the AER is specified by the boundary between *rfrng*-expressing dorsal and nonexpressing ventral ectoderm. FGF-8 from the AER maintains mesenchymal

FGF-10 expression and induces Shh expression in the ZPA region. Shh induces FGF-4 expression in the posterior AER. Expression of Shh and of FGF-4 are reciprocally maintained by a feedback loop between the ZPA and the posterior AER. At this stage, the limb region has budded out and proliferation of its mesechyme is maintained by FGF-8 and FGF-4 from the AER. Shh is required mainly for AP specification of digit number and identity (Litingtung et al. 2002); how the more proximal elements of the AP pattern are generated is unknown. Finally, the dorsal ectoderm expresses *Wnt-7*, which induces the transcription of *Lmx-1* in the dorsal mesoderm, which specifies dorsal. En-1 is expressed in the ventral ectoderm. It represses *Wnt-7* and thus *Lmx-1* expression there, resulting in a ventral pattern. There are many gaps in this model that need to be filled in by further investigation, but it can be used as a starting point to look at the specification of positional identity in the regenerating limb. What has been shown is that many of the same signaling molecules and transcription factors involved in patterning of the limb bud are also expressed in the regenerating limb and in similar (but not necessarily identical) temporal and spatial patterns.

6.4.1
Proximodistal Axis

The AEC is the distal signaling center of the regenerating amphibian limb, through its expression and (presumably) secretion of mitogens such as FGF-2 into the blastema. Outgrowth of the blastema and PD patterning are clearly linked, but the nature of the relationship between the cell cycle, mitogens supplied by the AEC and nerves, and distalization of PD positional identity is not understood. The *HoxA–D* gene clusters, which encode transcription factors, are known to be central mediators of axial patterning in vertebrates, including the amniote limb bud (McGinnis and Krumlauf 1992; Johnson and Tabin 1997). The expression patterns of these genes in regenerating limbs suggest that they are also involved in specifying forelimb vs. hindlimb and PD positional identity in regenerating limbs.

Two homeobox genes in newts associated with forelimb vs. hindlimb identity are *HoxC6*, which is expressed exclusively in forelimb blastemas, and *HoxC10*, which is expressed only in hindlimb blastemas (Tabin 1989; Simon and Tabin 1993). Homeobox genes that appear to play a role in patterning along the proximodistal axis of the regeneration blastema are *HoxD10* and *HoxA9* and *-A13*. The *HoxD/A* genes are expressed in 3′ to 5′ order within each *Hox* cluster (lower numbers to higher numbers), in overlapping domains from proximal to distal in the mesenchyme of developing limb buds and regeneration blastemas (Brown and Brockes 1991; Simon and Tabin 1993; Gardiner et al. 1995; Gardiner and Bryant 1996; Torok et al. 1998). Thus *HoxD10* and *-A9* expression is associated with proximal structures and *HoxD13* and *-A13* expression is associated with distal structures. *HoxD10* is expressed at a two- to threefold higher level in proximal vs. distal blastemas, and proximalization of distal blastemas by RA is associated with upregulation of *HoxD10* (Simon and Tabin 1993). *HoxA9* and *-A13* are both expressed from 1 day postamputation until redifferentiation begins in regenerating axolotl limbs (Gardiner et al. 1995). Each is expressed uniformly throughout the blastema until early bud, when expression of *HoxA13* starts to become restricted toward a more distal domain. The mechanism of this restriction is unknown. By the late bud stage, *HoxA13* expression is restricted to the region that will form the autopodium, suggesting that this gene is involved in patterning of autopodal structures. *HoxA13* expression in distal blastemas is 30% higher at the medium bud stage than in proximal blastemas. Proximalization of a distal blastema by RA reduces the expression of *HoxA13* by 55%, again suggesting that this gene is involved in distal patterning (Gardiner et al. 1995).

The expression of both proximal and distal *Hox* genes within 1 day postamputation, the subsequent restriction of distally expressed *Hox* genes to the prospective autopodium, and the reciprocal downregulation of distal genes and upregulation of proximal genes by RA, fits with the idea that regeneration in the PD axis is a process of self-organization via intercalary regeneration within distal and proximal boundaries that are present from the start of regeneration . We do not know whether *Hox/A9* and *-A13* are expressed in the same cells of the early blastema, or in different cells. If in different cells, a self-organizing PD axis might be set up by the sorting out of cells expressing the proximal and distal genes, with those cells expressing distal genes having an adhesive affinity for the AEC. If every cell is expressing both sets of genes, those cells in contact with the AEC might shut down the expression of *HoxA9* as a result of that association. In either case, distal cells would be juxtaposed to proximal cells, evoking intercalary regeneration of intermediate positional identities.

Notophthalmus viridescens (Nv)Tbox 1, a member of the T-box family of genes, is expressed in the mesenchyme of adult newt forelimb regeneration blastemas exclusively, suggesting that it, too, plays a role in specifying limb identity (Simon et al. 1997). Expression of *NvTbox1* is twofold higher in proximal vs. distal forelimb blastemas and RA increases expression of the gene twofold in distal blastemas, suggesting that it has a role in specifying PD positional identity as well (Simon et al. 1997).

6.4.2
Anteroposterior Axis

Hedgehog proteins play a key role in the patterning of a variety of embryonic systems. One member of the hedgehog family that is important for anteroposterior patterning in avian and mammalian limb buds is Sonic hedgehog (*Shh*), interacting in a feedback loop with FGF-4 in the apical ectoderm (Riddle et al. 1993; Johnson and Tabin 1997). In these limb buds, *Shh* is expressed from the earliest stages of bud formation in the ZPA. The ZPA evokes the duplication of digits when grafted anteriorly (Fallon and Crosby 1977). Cells transfected with the *Shh* gene and implanted into the anterior edge of chick limb buds also induce digit duplication (Riddle et al. 1993) and surgical removal of the *Shh*-expressing region results in defects of the posterior skeleton (Pagan et al. 1996), indicating that Shh plays a role in the polarizing activity of the ZPA.

Shh is expressed during the outgrowth of axolotl, newt, and *Xenopus* limb buds. Initially, expression is strong on the posterior margin of the bud, but then moves distally with outgrowth until it is restricted to the posterodistal region as digits begin to condense. Expression then ceases. *Shh* is also expressed in early to medium bud stages of regenerating newt, axolotl, and *Xenopus* tadpole limbs as well, but expression is lost in amputated *Xenopus* hindlimbs as the tadpoles approach metamorphosis, coincident with the proximal-to-distal loss of regeneration (Imokawa and Yoshizato 1997, 1998; Endo et al. 1997; Torok et al. 1999). The symmetrical cartilage spike that forms after amputation of regeneration-incompetent limbs result from a failure to establish a threshold circumference of normal neighbors in the absence of expression of *Shh* and/or *Hox* genes. When transfected into the anterior cells of axolotl blastemas, *Shh* evokes digit duplications (Roy et al. 2000). Two patches of *Shh* expression are seen after reversal of the AP axis of the newt limb regeneration blastema with respect to the rest of the limb, which evokes supernumerary limb regeneration. One is the original patch expressed by the posterior tissue of the primary blastema, which contributes to the supernumerary blastema on the anterior side of the limb. The other is a new patch that arises in the posterior edge of the new blastema evoked on the posterior side of the limb (Imokawa and Yoshizato 1997, 1998). A similar expression pattern is observed when supernumerary limb development is evoked by reversing the anteroposterior axis of young (stage 53) *Xenopus* limb bud tips (Endo et al. 1997).

Another hedgehog gene expressed in large areas of the early newt limb regeneration blastema is *Notophthalmus viridescens* banded hedgehog (*Nv-Bhh*) (Stark et al. 1998). *Bhh* is the equivalent of Indian hedgehog (*Ihh*) in other vertebrates (Zardoya et al. 1996). Ihh is known to be essential in signaling mechanisms that control cartilage proliferation, hypertrophic differentiation, and ossification in developing limbs (Vortkamp 2001), but its very early appearance during regeneration implies that it has other (as yet unknown) roles.

The function of hedgehog proteins in patterning of the regeneration blastema is not yet clear, however. The relatively late expression of the *Shh* gene (at medium bud) may mean that it is not required for AP patterning of most of the limb bud or regeneration blastema, but rather is involved in specification of digit identity. Genetic analysis of patterning in the mouse limb bud in fact indicates that the skeletal elements of the stylopodium and zeugopodium develop normally in the absence of *Shh*. Digits also develop in the absence of *Shh*, but they are iterative, so *Shh* is thought to be required for normal digit number and identity (Litingtung et al. 2002).

6.4.3 Dorsoventral Axis

In embryonic amniote limb buds, the dorsal ectoderm is thought to be a signaling center controlling DV pattern (Akita 1996). DV reversal of the early chick limb bud ectoderm with respect to the underlying mesoderm reverses the DV axial polarity of the mesoderm, to conform with that of the ectoderm. The *Wnt-7a* gene is expressed in the dorsal ectoderm and in turn induces expression of the gene *Lmx-1* in the dorsal mesoderm. Dorsal limb mesoderm is ventralized by loss of *Wnt-7a* or *Lmx-1* function; furthermore, misexpression of these genes in ventral mesoderm converts it to a dorsal fate (Parr and McMahon 1995; Cygan et al. 1997;

Chen et al. 1998; Rodriguez-Esteban et al. 1998). Conversely, the gene *En-1* is expressed in the ventral ectoderm. Loss of its function leads to expression of *Wnt-7a* in the ventral ectoderm and dorsal pattern duplication (Cygan et al. 1997; Loomis et al. 1998; Christen and Slack 1998). Ectopic expression of *En-1*, however, does not result in dorsalization of mesoderm (Logan et al. 1997). These results suggest that *En-1* normally represses *Wnt-7a* expression in ventral ectoderm, and that the antagonistic effects of these two genes is the basis of DV polarity.

Investigations of gene expression associated with DV axial polarity during limb regeneration have so far been carried out only in *Xenopus*. *Xenopus Lmx-1* was cloned and found to be expressed in the dorsal mesenchyme of limb buds at stages 51–53 (regeneration-competent) and of regeneration blastemas formed after amputation through the zeugopodium at these stages. Expression was not seen in stage 55 blastemas, which formed only symmetrical cartilage spikes (Matsuda et al. 2001). The normal spatial relationship between the epidermis and mesodermal tissues was reversed by stripping the epidermis from the zeugopodial segment of the limb bud and reversing the DV axis of this segment with respect to the limb stump. Fresh epidermis of normal DV polarity grew over the reversed segment, which was then allowed to regenerate. When this operation was performed at stage 52, *Lmx-1* expression and structural pattern of the regenerate conformed to the epidermal polarity; i.e., was reversed. But when the operation was performed at stage 55, the original *Lmx-1* expression and structural polarity of the segment were maintained in the regenerate (Matsuda et al. 2001). These results suggest that self-organization of DV polarity in the limb regeneration blastema is specified by the effect of the wound epidermis on the expression of *Lmx-1* up through stage 52 or 53; after that the polarity is fixed. The expression patterns of *Wnt-7a* and *En-1* in the regeneration blastema have not yet been reported.

7
Other Regenerating Systems

Larval and adult salamanders and newts can regenerate the upper and lower jaws (Goss and Stagg 1958; Graver 1973; Ferretti and Ghosh 1997) and newts can regenerate the intestine (Goodchild 1956; O'Steen and Walker 1962; Grubb 1975) by forming blastemas of undifferentiated cells that proliferate and replace the missing parts. The dedifferentiated cells for jaw regeneration are derived from jaw muscle, bone, cartilage, and connective tissue fibroblasts; those for intestinal regeneration appear to be derived from serosal and smooth muscle cells. Newts can also regenerate cardiac muscle after amputation of the tip of the ventricle by the partial dedifferentiation and proliferation of cardiomyocytes (Oberpriller and Oberpriller 1991). The amount of new cardiac muscle regenerated, however, is low and mixed with fibroblastic scar tissue. Higher rates of proliferation and amounts of regenerated cardiac muscle were obtained by mincing the amputated tip of the ventricle and placing the mince on the wound surface (Bader and Oberpriller 1978). The mechanisms of regeneration in these systems have not been well studied as yet. However, the mechanisms of regeneration in two other systems, the lens and the neural retina (NR), have been investigated in some depth.

The embryos of several vertebrate species, including fish, frogs, birds, and rats, the tadpoles of the anuran, *Xenopus laevis*, and the adult newt, can regenerate the lens and NR (Reyer 1977; Stroeva and Mitashov 1983; Park and Hollenberg 1993; Mitashov 1996). The newt regenerates the lens from pigmented epithelial cells of the dorsal iris after lentectomy, or, if the dorsal iris is missing, from the retinal pigmented epithelium (RPE) (Reyer 1977; Stroeva and Mitashov 1983; Mitashov 1996; Eguchi 1998). *Xenopus* tadpoles regenerate the lens from the inner corneal epithelium (Freeman 1963). Cutting the optic nerve and blood vessels to the eye in adult newts and anuran tadpoles results in degeneration of the NR, followed by its regeneration, along with the optic nerve, from dedifferentiated RPE and ciliary body (ora serrata) cells (Sperry 1944; Gaze 1959; Mitashov 1996; Raymond and Hitchcock 1997). The dedifferentiated pigmented epithelial cells proliferate and transdifferentiate into lens cells or the cells that compose the different layers of the NR: ganglion cells, inner and outer nuclear cells, and photoreceptor cells.

7.1 Neural Retina Regeneration

Members of the FGF family play a prominent role in regeneration of the NR and lens, as they do in limb regeneration. Studies on chick embryos suggest that FGF-2 is an important signaling molecule in NR regeneration. FGF-2 is expressed in the eye tissues of birds and mammals (Noji et al. 1990; Park and Hollenberg 1993). During optic cup formation in stage 9–10 embryos, the prospective RPE (outer layer of the optic cup) differentiates as NR if its contact with the prospective NR (inner layer of the optic cup) is broken (Orts-Lorca and Genis-Galvez 1960), suggesting that the prospective NR normally inhibits the prospective RPE from differentiating as NR. By stage 22–24, the RPE and NR are differentiated and the RPE will not form NR when the NR is completely removed. However, the RPE can be stimulated to regenerate the NR by implanting a piece of NR back into the vitreous chamber (Coulombre and Coulombre 1965, 1970) or by implanting blocks of ethylene/vinyl coacetate polymer impregnated with 10 ng of FGF-2, or, at higher concentrations, FGF-1 (Park and Hollenberg 1991). In both cases, the polarity of the regenerated NR is reversed, with the photoreceptor layer facing the implant, implying that RPE cells experiencing the highest concentration of FGF become photoreceptors. Incubation of stage 9–10 chick optic vesicles in medium containing FGF-2 also induced the prospective RPE to differentiate as NR, producing an eye with two NRs of normal polarity (Pittack et al. 1997). FGF-2 has also been shown to induce RPE cells to form NR cells in vitro (Pittack et al. 1991).

Collectively, these results suggest that, during chick embryogenesis, the RPE and NR exert a reciprocal inhibitory effect on one another. The NR prevents the RPE from becoming NR through as yet unidentified signals, while the RPE prevents the NR from producing FGF-2. If the NR is separated from the RPE or removed and a piece put back, it produces FGF-2, which stimulates RPE cells to regenerate NR. Since in amphibians the RPE regenerates the NR after complete destruction of the original NR, either no FGF signal is required for regeneration, or enough FGF is released from the degenerating NR, or the RPE is released from an inhibitory influence of the NR and itself produces FGF, which acts in an autocrine fashion. Although FGF receptors are expressed in the intact retina of the adult newt (see below), there are no data on expression of FGFs or receptors in the regenerating amphibian neural retina.

7.2 Lens Regeneration

Several lines of evidence suggest that the lens regeneration-promoting factor produced by the NR may be FGF-1. FGF-1 transcripts are expressed in the intact eye and are upregulated in regenerating lens cells of newts (Del-Rio Tsonis et al. 1997). FGF-1 stimulates cultured newt dorsal iris and *Xenopus* corneal cells to transdifferentiate into lens cells, though this transformation is not associated with the formation of a normally organized lens (Bosco et al. 1993, 1997). FGF-2 does not have this effect in vivo (Del-Rio Tsonis et al. 1995), although it can induce the transdifferentiation of chick RPE cells to lens cells in vitro (Mochii et al. 1998). Exogenous FGF-1, which binds to FGFR-1 and to the KGFR variant of FGFR-2, and FGF-4, which binds to the *bek* variant of FGFR-2, cause structural abnormalities in regenerating newt lenses when applied via implanted beads, a result similar to that seen in the developing lenses of FGF transgenic mice (Del-Rio Tsonis 1997).

In situ hybridization experiments have shown that transcripts for FGFR-1 are expressed in the NR and lens epithelium of the intact adult newt eye. After lentectomy, expression of the FGFR-1 gene remains the same in the retina, but is upregulated during dedifferentiation of dorsal iris cells and throughout their transdifferentiation into lens cells (Del-Rio Tsonis et al. 1998). Transcripts are also expressed in the ciliary body and the ventral iris, though at a lower level than in the dorsal iris. A similar pattern of expression of FGFR-2 and FGFR-3 is observed in the intact eye and regenerating lens (Del-Rio Tsonis et al. 1997). Despite the widespread expression of FGFR-1 transcripts, immunostaining showed that the FGFR-1 protein is confined to the dedifferentiating cells of the newt dorsal iris that form the lens, indicating that expression of the gene is regulated at the level of translation, as well as transcription (Del-Rio Tsonis et al. 1998). A further indication of the central importance of FGFR-1 (and FGF-1) is that lens regeneration is inhibited by SU5402, a 3-substituted indolin-2-one that specifically inhibits the autophosphorylation, and thus activation, of this receptor (Del-Rio Tsonis et al. 1998). The protein expression patterns of FGFR-2 and 3, as well as FGFR-1, have also been examined by immunoblotting (McDevitt et al. 1997). They reported no regional, temporal, qualitative or quantitative differences for FGFR-1 and 2 expression. A higher level of FGFR-3, however, was observed in intact dorsal iris and in dorsal iris cells as they engage in lens regeneration.

Eye development, like the limb, is highly dependent on RA signaling through its nuclear receptors to activate the α B crystallin gene. RA signaling is dependent on the transcription factor Pax-6, as shown by the fact that in mice mutated for *Pax-6*, RA signaling in the eye is decreased and the developing lens cannot respond to exogenous RA (Enwright and Grainger 2000). Inhibition of RA synthesis by disulfram, or RAR function by the RAR antagonist 193109, resulted in the inhibition, retardation, or abnormal morphogenesis of regeneration, although in some cases ectopic lenses regenerated from the ventral iris or the cornea (Tsonis et al. 2000). Collectively, these observations suggest that RARs and Pax-6 act together in promoting the dedifferentiation and transdifferentiation of dorsal iris cells to lens cells, but that elsewhere in the eye (ventral iris and cornea) RARs function to inhibit ectopic lens regeneration. RAR-δ, which mediates PD positional effects of RA in regenerating newt limbs (Pecorino et al. 1996), may also mediate the effects of RA on lens regeneration. RAR-δ transcripts are expressed at low levels in the ganglion cell layer of the retina, but not elsewhere in the eye. After lentectomy the RAR-δ gene is expressed in the dedifferentiated cells that form the new lens vesicle, and the level of expression increases to its highest level during lens fiber differentiation (Tsonis et al. 2000).

8 Relevance of Amphibian Regeneration to Regenerative Medicine

Three major strategies are being developed for use as regenerative medicine. The first two are transplantation of cells as aggregates or suspensions, and construction of bioartificial tissues by combining cells with a biomimetic scaffold that degrades after implanting the construct into the body. The cells to be used are precursor cells differentiated from ESCs or adult stem cells harvested from the patient. Each of these has its advantages and limitations (Stocum 2001). ESCs can give rise to teratocarcinomas (Evans and Kaufman 1983) and thus must be differentiated to an adult stem cell level before transplantation. Directing this differentiation has proved difficult. ESCs exhibit an epigenetic instability that could be problematic as part of a functional tissue (Humphries et al. 2001). Furthermore, ESC derivatives are subject to immunorejection. Although this can be overcome by somatic nuclear transfer to make 'personal' ESC lines, such procedures raise bioethical issues. Nevertheless, as we understand more about the signals that direct embryonic tissue development, progress is being made in the directed differentiation of ESC derivatives to correct tissue deficiencies. Multipotential glial precursor cells, derived by the directed differentiation of mouse ESCs in vitro, differentiated into astrocytes and oligodendrocytes which remyelinated spinal cord and brain axons when injected into mutant rats suffering from a myelin-deficiency condition that mimics human Pelizaeus–Merzbacher disease (Brustle et al. 1999). Functional dopaminergic neurons differentiated from mouse ESCs and grafted into the striatum of animals lesioned by administration of 6-hydroxy dopamine, integrated into the host tissue and reversed the Parkinsonian-like symptoms created by this treatment (Kim et al. 2002).

The use of adult stem cells as regenerative therapies is based on their plasticity and lack of bioethical controversy, since they would be derived from the patient's own body. There is some skepticism, however, as to whether adult reserve stem cells are as plastic as has been claimed. The level of lineage switching of these cells is low when injected into adult animals and some studies have not been repeatable. Furthermore, it is possible that what appears to be lineage-switching is actually fusion of donor cells with host cells; such fusion has been demonstrated in vitro (Terada et al. 2002; Ying et al. 2002). Finally, functionality of multiple cell types derived from a single adult stem cell has not been demonstrated in most cases. However, a much more convincing case for the potential of adult stem cells in regenerative medicine has been made with the recent demonstration of a multipotential adult precursor cell (MAPC) in human and mouse bone marrow that clonally gives rise to functional derivatives of all three germ layers in vitro (human) or in vivo (mouse) (Reyes et al. 2001, 2002; Schwartz et al. 2002; Jiang et al. 2002). A major question that needs to be answered with regard to the use of adult stem cells is whether the known decline in regenerative ability with age is due to a decline in the quantity or quality (proliferative and differentiative potential) of adult stem cells, to a diminished signaling environment, or both.

The third approach of regenerative medicine is the induction of regeneration at the site of tissue damage by stimulating cells that are competent to regenerate cell parts (for example, the axons of neurons) or proliferate (reserve stem cells). This is the most desirable way to regenerate a tissue, because it relies strictly on autogeneic cells activated at the site of damage, requires less invasive procedures, and the regenerated tissue is more likely to be well-integrated with the old tissue. Presumably, the cells that are activated would be reserve adult stem cells. These cells could be local or be attracted from elsewhere in the body, such as MAPCs from the bone marrow, where they would respond to local signals, or to exogenously supplied signals (inductors), by differentiating into the appropriate cell types. This may not be as easy as it seems, however. We do not know how ubiquitous reserve stem cells are, though they have been found in more tissues than previously believed (Stocum

2001). Inducing bone marrow or other stem cells to make a variety of different cell types may not be straightforward. For example, the MAPCs described by Jiang et al (2002) were able to accomplish this trick only after extensive division in vitro.

Another possibility for the induction of regeneration is to stimulate the dedifferentiation of cells at an injury site. Since the amphibians are the divas of dedifferentiation, they are valuable experimental models for understanding how to accomplish this feat. We would like to be able to regenerate not just tissues or parts of tissues, but complex structures such as the limb, spinal cord, and neural retina that require precise spatial patterning to be functional. The factors involved in patterning can be readily investigated in the regeneration of complex structures such as the amphibian limb.

A huge advantage to incorporating research on amphibian regeneration into regenerative biology and medicine is that cellular and molecular comparisons between regenerating and nonregenerating tissues can be made to discover the patterns of gene activity that distinguish regeneration from scar tissue formation, both in terms of gene products that promote regeneration, but equally important, gene products that inhibit regeneration and thus must be neutralized to induce it. Such comparisons can be made between amphibians that do and do not regenerate (for example, axolotl vs. adult frog), between regeneration-competent and -deficient developmental stages of amphibians (tadpoles vs. metamorphosed frogs) or between regenerating amphibian and nonregenerating mammalian tissues (axolotl or newt vs. mouse).

Such studies on are underway in a number of laboratories. For example, a recent set of experiments suggests that mammalian cells have the latent capacity to dedifferentiate. Mouse myonuclei will enter S in response to serum stimulation if they are part of a heterokaryon made by fusing C2C12 and newt myoblasts, indicating that the intracellular pathways to Rb phosphorylation are the same in newt and mouse (Velloso et al. 2001). McGann et al (2001) have shown that cultured newt myotubes respond to a protein extract of regenerating newt limbs by cleaving into mononucleate cells and dedifferentiating. Cultured C2C12 mouse myotubes, which do not normally dedifferentiate, also cleave into mononucleate cells, undergo morphological dedifferentiaton, and proliferate in response to the newt extract. Expression of muscle regulatory proteins is reduced to undetectable levels in these cells. Control myotubes exposed to a protein extract from unamputated limbs do not dedifferentiate. The extract of regenerating newt limbs clearly harbors factors in addition to just the thrombin-activated protein (Tanaka et al. 1999), that allow muscle cells to fragment and lose their morphological and molecular phenotypic specializations. The isolation and characterization of these factors is thus of high importance.

9 Summary and Conclusions

Amphibians regenerate the limb and other complex structures by creating stem cells through a process of histolysis and dedifferentiation of mature cells. These cells re-enter the cell cycle, proliferate and differentiate into the parts lost by injury. Muscle is the most complex cell type that dedifferentiates. Serum contains a thrombin-activated factor, distinct from the standard growth factors found in serum, that drives the nuclei of newt myofibers into S phase, but completion of the cell cycle requires fragmentation of the myofibers into single cells. Proliferation of dedifferentiated cells requires mitogens supplied by the AEC and the limb nerves. Neural molecules that satisfy one or more of the criteria for mitogen status are FGF-2, GGF-2, substance P, and transferrin. Only transferrin has been tested extensively enough to determine that it satisfies all of the criteria. AEC mitogen candidates are FGF-1 and -2.

The blastema is a self-organizing system, within which individual stem cells exhibit a limited developmental plasticity. The cellular and molecular basis of self-organization is not completely understood. One model consistent with the experimental data is based on the inheritance of boundary positional identities from parent limb cells within which other positional identities are intercalated until a normal neighbor map is achieved. Circumferential blastema cells act as boundaries for intercalation of transverse axial pattern along the radii of the circumference. Some of the circumferential cells have the distal-most positional identity and others the proximal-most; the PD axial pattern is intercalated between them. The blastema expresses some of the same kinds of signaling molecules and transcription factors, in similar temporal and spatial patterns, as the three signaling centers of the amniote limb bud. While there are differences between limb buds and regeneration blastemas in the source of the cells used to construct the limb parts, the basic patterning mechanisms appear to have much in common.

The regenerating limb and other systems in amphibians are valuable models for the study of mechanisms of regeneration of complex structures and how to apply the knowledge gained to the induction of regeneration in mammals. For example, mammalian muscle cells have the latent ability to fragment, dedifferentiate, and proliferate in response to factors present in a protein extract of regenerating newt limbs. Obtaining a molecular inventory of molecules that stimulate regeneration in amphibians, as well as molecules that inhibit regeneration in mammals, will offer the possibility of devising a cocktail of stimulatory proteins plus neutralizers of inhibitory proteins for use in regenerative therapies.

References

Akita K (1996) The effect of the ectoderm on the dorsoverntral pattern of epidermis, muscles and joints in the developing chick leg: a new model. Anat Embryol (Berlin) 193:377–386

Albert P, Boilly B, Courty J, Barritault D (1987) Stimulation in cell culture of mesenchymal cells of newt limb blastemas by EDGFI Or II (basic or acidic FGF). Cell Diff 21:63–68

Albert P, Boilly B (1988) Effect of transferrin on amphibian limb regeneration: a blastema cell culture study. Roux's Archiv Dev Biol 197:193–196

Alvarado AS (2000) Regeneration in the metazoans: why does it happen? Bioessays 22:578–590

Bader D, Oberpriller JO (1978) Repair and reorganization of minced cardiac muscle in the adult newt (*Notophthalmus viridescens*). J Morph 155:349–358

Blau HM, Brazelton TR, Weimann JM (2001) The evolving concept of a stem cell: entity or function. Cell: 105:829–841

Boilly B, Cavanaugh KP, Thomas D, Hondermarck H, Bryant SV, Bradshaw RA (1991) Acidic fibroblast growth factor is present in regenerating limb blastemas of axolotls and bonds specifically to blastema tissues. Dev Biol 145:302–310

Bosco, L., Valle, C, and Willems, D (1993) In vivo and in vitro experimental analysis of lens regeneration in larval *Xenopus laevis*. Dev. Growth Diff 35:257–270

Bosco, L., Venturini, G., and Willems, D (1997) In vitro lens transdifferentiation of *Xenopus laevis* outer cornea induced by Fibroblast Growth Factor (FGF). Development 124:421–428

Brockes JP (1984) Mitogenic growth factors and nerve dependence of limb regeneration. Science 225:1280–1287

Brockes JP (1992) Introduction of a retinoid reporter gene into the urodele limb blastema. Proc Natl Acad Sci UA 89:11386–11390

Brockes JP (1997) Amphibian limb regeneration: rebuilding a complex structure. Science 276:81–87

Brockes JP, Kintner CR (1986) Glial growth factor and nerve-dependent proliferation in the regeneration blastema of urodele amphibians. Cell 45:301–306

Brown R, Brockes JP (1991) Identification and expression of a regeneration-specific homeobox gene in the newt limb blastema. Development 111:489–496

Brustle O, Jones KN, Learish RD, Karram K, Choudhary K, Wiestler OD, Duncan ID, McKay RD (1999) Embryonic stem cell-derived glial precursors. A source of myelinating transplants. Science 285:754–756

Bryant V, Iten LE (1976) Supernumerry limbs in amphibians: Experimental production in *Notophthalmus viridescens* and a new interpretation of their formation. Dev Biol 50:212–234

Bryant S, French V, Bryant PJ (1981) Distal regeneration and symmetry. Science 212:993–1002

Cadinouche MZA, Liversage RA, Muller W, Tsilfidis C (1999) Molecular cloning of the *Notophthalmus viridescens Radical Fringe* cDNA and characterization of its expression during forelimb development and adult forelimb regeneration. Dev Dyn 214:259–268

Cameron JA, Hilgers AR, Hinterberger TJ (1986) Evidence that reserve cells are a source of regenerated adult newt muscle in vitro. Nature 321:607–610

Cameron JA, Fallon JF (1977 Evidence for polarizing zone in the limb buds of *Xenopus laevis*. Dev Biol 55:320–330

Cannata M, Bernardini S, Filoni S (1992) Regenerative responses in cultured hindlimb stumps of larval *Xenopus laevis*. J exp Zool 262:446–453

Carlson BM (1974) Morphogenetic interactions between rotated skin cuffs and underlying stump tissues in regenerating axolotl forelimbs. Dev Biol 39:263–285

Carlson BM (1975) The effects of rotation and positional change of stump tissues upon morphogenesis of the regenerating axolotl limb. Ev Biol 47:269–291

Carlson MRJ, Bryant SV, Gardiner DM (1998) Expression of *Msx-2* during development, regeneration, and wound healing in axolotl limbs. J Exp Zool 282:715–723

Casimir CM, Gates P, Patient RK, Brockes JP (1988) Evidence for dedifferentiation and metaplasia in amphibian limb regeneration from inheritance of DNA methylation. Development 104:657–668

Chapron C (1974) Mise en evidence du role, dans la regeneration des amphibians, d'une glycoproteine secretee par la cape apicale: Etude cytochemique st autoradiographique en microscopie electronique. J Embryol exp Morph 32:133–145

Chen H, Lun Y, Ovchinnikov D, Kokubo H, Oberg KC, Pepicelli CV, Gan L, Lee B, Johnson RL (1998 Limb and kidney defects in *Lmx1b* mutant mice suggest and involvement of LMX1B in human nail patella syndrome. Nature Genet 19:51–55

Chernoff, EAG (1996 Spinal cord regeneration: a phenomenon unique to urodeles? Int J Dev Biol 40:823–832

Chew K, Cameron JA (1983) Increase in mitotic activity of regenerating axolotl limbs by growth factor-impregnated implants. J exp Zool 226:325–329

Christen B, Slack JMW (1997) *FGF-8* is associated with anteroposterior patterning and limb regeneration in *Xenopus*. Dev Biol 192:455–466

Christen B, Slack JMW (1998) All limbs are not the same. Nature 395:230–231

Christensen RN, Tassava RA (2000) Apical epithelial cap morphology and fibronectin gene expression in regenerating axolotl limbs. Dev Dyn 217:216–224

Christensen RN, Weinstein M, Tassava RA (2001) Fibroblast growth factors in regenerating limbs of *Ambystoma*: cloning and semi-quantitative RT-PCR expression studies. J exp Zool 290:529–540

Christensen RN, Weinstein M, Tassava RA (2002) Expression of fibroblast growth factors 4,8,and 10 in limbs, flanks, and blastemas of *Ambystoma*. Dev Dyn 223:193–203

Clark RAF (1996) Wound repair: Overview and general considerations. In: Clark RAF (ed) The molecular and cellular biology of wound repair. John Wiley & Sons, New York, pp 255–272

Clarke D, Johansson C, Wilbertz J, Veress B, Nilsson E, Karlstrom H, Lendahl U, Frisen J (2000) Generalized potential of adult neural stem cells. Science 288:1660–1663

Coulombre JL, Columbre AJ (1965) Regeneration of neural retina from the pigmented epithelium in the chick embryo. Dev Biol 12:79–92

Coulombre JL Coulombre AJ (1970) Influence of mouse neural retina on regeneration of chick neural retina from chick embryonic pigmented epithelium. Nature 228:559–560

Crawford K, Stocum DL (1988a) Retinoic acid coordinately proximalizes regenerate pattern and blastema differential affinity in axolotl limbs. Development 102:687–698

Crawford K, Stocum DL (1988b) Retinoic acid proximalizes level-specific properties responsible for intercalary regeneration in axolotl limbs. Development 104:703–712

Crews L, Gates PB, Brown R, Joliot A, Foley C, Brockes, J, Gann A (1995) Expression and activity of the newt Msx-1 gene in relation to limb regeneration. Proc Royal Soc London [Biol] 259:161–171

Cygan JA, Johnson L, McMahon AP (1997) Novel regulatory interactions revealed by studies of murine limb pattern in *Wnt-7a* and *En-1* mutants. Development 124:5021–5032

Del-Rio Tsonis K, Washabaugh CH, Tsonis PA (1995) Expression of *pax-6* during urodele eye development and lens regeneration. Proc Natl Acad Sci USA 92:5092–5096

DelRio Tsonis K, Jung JC, Chiu, I-M, and Tsonis P.A. 1997 Conservation of fibroblast growth factor function in lens regeneration. Proc Natl Acad Sci USA 94:13701–13706

Del-Rio Tsonis K, Trombley MT, McMahon G, Tsonis PA (1998) Regulation of lens regeneration by fibroblast growth factor receptor 1. Dev Dyn 213:140–146

Dent JN (1962) Limb regeneration in larvae and metamorphosing individuals of the South African clawed toad. J Morph 110:61–77

D'Jamoos CA, McMahon G, Tsonis PA (1998) Fibroblast growth factor receptors regulate the ability for hindlimb regeneration in *Xenopus laevis*. Wound Rep Reg 6:388–397

Desselle J-C, Gontcharoff M (1978) Cytophotometric detection of the participation of cartilage grafts in regeneration of x-rayed urodele limbs. Biol Cellul 33:45–54

Dresden MH (1969) Denervation effects on newt limb regeneration: DNA, RNA, and protein synthesis. Dev Biol 19:311–320

Dresden MH, Gross J (1970) The collagenolytic enzyme of the regenerating limb of the newt *Triturus viridescens*. Dev Biol 22:129–137

Ducy P (2000) Cbfa1: A molecular switch in osteoblast biology. Dev Dyn 219:461–471

Ducy P, Schinke T, Karsenty G (2000) The osteoblast: A sophisticated fibroblast under central surveillance. Science 289:1501–1504

Dudley AT, Ros MA, Tabin CJ (2002) A re-examination of proximodistal patterning during vertebrate limb development. Nature 418:539–544

Dungan KM, Wei TY, Nace JD, Poulin ML, Chiu I-M, Lang JC, Tassava RA (2002) Expression and biological effect of urodele fibroblast growth factor 1: Relationship to limb regeneration. J exp Zool 292:540–554

Dunis DA, Namenwirth M (1977) The role of grafted skin in the regeneration of X-irradiated axolotl limbs. Dev Biol 56:97–109

Echeverri K, Clarke DW, Tanaka EM (2001) In vivo imaging indicates muscle fiber dedifferentiation is a major contributor to the regenerating tail blastema. Dev Biol 236:151–164

Egar MW (1993) Affinophoresis as a test of axolotl accessory limbs. In: Fallon JF, Goetinck PF, Kelley RO, Stocum DL (eds) Limb development and regeneration, Part B. Wiley-Liss, New York pp. 203–211

Egar M, Singer M. (1981) The role of ependyma in spinal cord regrowth. In: Becker RO (ed) Mechanisms of growth control. Charles Thomas Publisher, Springfield pp 93–106

Eguchi, G (1998) Transdifferentiation as the basis of eye lens regeneration. In: Cellular and molecular basis of regeneration. Ferretti P, Geraudie J (eds) John Wiley & Sons, New York pp 207–229

Einhorn TA (1998) The cell and molecular biology of fracture healing. Clin Orthopaed Related Res 355S:7–21

Endo T, Yokoyama H, Tamura, Ide H (1997) *Shh* expression in in developing and regenerating limb buds of *Xenopus laevis*. Dev Dyn 209:227–232

Enwright JF, Grainger RM. (2000) Altered retinoid signaling in the heads of small eye mouse embryos. Dev Biol. 221:10–22

Evans MJ, Kaufman M (1983) Pluripotential cells grown directly from normal mouse embryos. Cancer Surv 2:185–208

Fallon JF, Crosby GM (1977) Polarizing zone activity in limb buds of amniotes. In: Ede DA, Hinchliffe JR, Balls M, eds. Vertebrate Limb and Somite Morphogenesis. Cambridge University Press, Cambridge pp 55–70

Fallon JF, Lopez A, Ros, MA, Savage MP, Olwin BB, Simandl BK (1994) Apical ectodermal ridge growth signal for chick limb development. Science 264:104–107

Fekete D, Brockes JP (1987) A monoclonal antibody detects a difference in the cellular composition of developing and regenerating limbs of newts. Development 99:589–602

Fekete DM, Brockes JP (1988) Evidence that the nerve controls molecular identity of progenitor cells for limb regeneration. Development 103:567–573

Ferretti P, Brockes JP (1991) Cell origin and identity in limb regeneration and development. Glia 4:214–224

Ferretti P, Ghosh S (1997) Expression of regeneration-associated cytoskeletal proteins reveals differences and similarities between regenerating organs. Dev Dyn 210:288–304

Feynman R, Leighton R, Sands M (1977a) The Feynman lectures on physics, Vol II. Addison-Wesley Publishing Company, New York, pp 19–4 to 19–14

Feynman RP, Leighton RB, Sands M (1977b) The Feynman lectures on physics, Vol I. Addison-Wesley Publishing Company, New York, pp 26-3 to 26-8

Filoni S, Paglialunga L (1990) Effect of denervation on hindlimb regeneration in *Xenopus laevis* larvae. Differentiation 43:10–19

Filoni S, Bernardini S, Cannata SM (1991) The influence of denervation on grafted hindlimb regeneration of larval *Xenopus laevis*. J exp Zool 260:210–219

Filoni S, Bernardini S, Cannata SM, Ghittoni R (1999) Nerve-independence of limb regeneration in larval *Xenopus laevis* is related to the presence of mitogenic factors in early limb tissues. J exp Zool 284:188–196

Freeman G (1963) Lens regeneration from cornea in *Xenopus laevis*. J exp Zool. 154:39–65

French V, Bryant PJ, Bryant SV (1976) Pattern regulation in epimorphic fields. Science 193:969–981

Fuchs E, Segre JA (2000) Stem cells: a new lease on life. Cell:143–156

Gardiner DM, Blumberg B, Komine Y, Bryant SV (1995) Regulation of *HoxA* expression in developing and regenerating axolotl limbs. Development 121:1731–1741

Gardiner DM, Bryant SV (1996) Molecular mechanisms in the control of limb regeneration: the role of homeobox genes. Int J Dev Biol 40:797–805

Gaze RM. (1959) Regeneration of the optic nerve in Xenopus laevis. Quart J exp Physiol. 44:290–308

Geraudie J, Ferretti P (1998) Gene expression during amphibian limb regeneration. Int J Cytol 180:1–50

Gevers W (1984) Protein metabolism in the heart. J Mol Cell Cardiol 16:3–32

Giguere V, Ong S, Evans RM, Tabin CJ (1989) Spatial and temporal expression of the retinoic acid receptor in the regenerating amphibian limb. Nature 337:566–569

Gilbert SF (2000) Developmental biology. Sinauer Associates, Inc, Sunderland MA

Globus M (1988) A neuromitogenic role for substance P in urodele limb regeneration. In Inoue S, Shirai T, Egar M, Aiyama S, Geraudie J, Nobunaga T, Sato NL (eds). Regeneration and development. Okada Printing & Publishing, Maebashi, pp 675–685

Globus M, Alles P (1990) A search for immunoreactive substance P and other neural peptides in the limb regenerate of the newt *Notophthalmus viridescens*. J exp Zool 254165–176

Globus M, Vethamany-Globus S, Lee YCI (1980) Effect of apical epidermal cap on mitotic cycle and cartilage differentiation in regeneration blastemata in the newt, *Notophthalmus viridescens*. Dev Biol 75:358–372

Globus M, Vethamany-Globus S (1985) In vitro studies of controlling factors in newt limb regeneration. In: Sicard RE (ed) Regulation of Vertebrate Limb Regeneration. New York, Oxford University Press. pp 106–127

Go MJ, Eastman DS, Artavanis-Tsakonas S (1998) Cell proliferation control by Notch signaling in *Drosophila* development. Development 125:2031–2040

Goldhamer DJ, Tassava RA (1987) An analysis of proliferative activity in innervated and denervated forelimb regenerates of the newt *Notophthalmus viridescens*. Development 100:619–628

Goodchild CG (1956) Reconstitution of the intestinal tract in the adult leopard frog *Rana pipiens Schreber*. J exp Zool 131:301–327

Goss RJ (1956a Regenerative inhibition following limb amputation and immediate insertion into the body cavity. Anat Rec 126:15–27
Goss RJ (1956b) The regenerative responses of amputated limbs to delayed insertion into the body cavity. Anat Rec 126:283–297
Goss RJ (1969) Principles of regeneration. Academic Press, New York
Goss RJ, Stagg MW (1958) Regeneration of lower jaws in adult newts. J Morph 102:289–310
Goss RJ, Holt R (1992) Epimorphic vs tissue regeneration in *Xenopus* forelimbs. J exp Zool 261451–457
Graver H (1973) The polarity of the dental lamina in the regenerating salamander jaw. J Embryol exp Morph 30:635–646
Grillo H, Lapiere CM, Dresden MH, Gross J (1968) Collagenolytic activity in regenerating forelimbs of the adult newt (*Triturus viridescens*). Dev Biol 17:571–583
Grubb RB (1975) An autoradiographic study of the origin of intestinal blastema cels in the newt *Notophthalmus viridescens*. Dev Biol 47:185–195
Gussoni E, Soneoka Y, Strickland CD, Buzney EA, Khan MK, Flint AF, Kunkel LM, Mulligan RC (1999) Dystrophin expression in the mdx mouse restored by stem cell transplantation. Nature 401:390–394
Guyenot E (1927) La perte du pouvoir regenerateur des Anoures, etudiee par les heterogreffess, et la notion de territories. Rev Suisse Zool 34:1–54
Han M-J, An J-Y, Kim W-S (2001) Expression patterns of *Fgf-8* during development and limb regeneration of the axolotl. Dev Dyn 220:40–48
Hay ED (1959) Electron microscopic observations of muscle dedifferentiation in regenerating *Amblystoma* limbs. Dev Biol 3:26–59
Hill RE, Jones PF, Rees AR, Sime CM, Justice MJ, Copeland NG, Jenkins NA, Graham E, Davidson DR (1989) A new family of mouse homeobox-containing genes: molecular structure, chromosomal location, and developmental expression of Hox-7.1. Genes Dev 3:26–37
Hill DS, Ragsdale CW, Brockes JP (1993) Isoform-specific immunological detection of newt retinoic acid receptor $\delta 1$ in normal and regenerating limbs. Development 117:937–945
Holder, N (1989) Organization of connective tissue patterns by dermal fibroblasts in the regenerating axolotl limb. Development 105:585–593
Holder N, Tank PW (1979) Morphogenetic interactions occurring between blastemas and stumps after exchanging blastemas between normal and double-half forelimbs in the axolotl *Ambystoma mexicanum*. Dev Biol 74:302–314
Humpherys D, Eggan K, Akutsu H, Hochedlinger K, Rideout WM III, Biniszkiewicz D, Yanagimachi R, Jaenisch R (2001) Epigenetic instability in ES cells and cloned mice. Science 293:95–97
Ide H, Wada N, Uchiyama K (1994) Sorting out of cells from different parts and stages of the chick limb bud. Dev Biol 162:71–76
Ide H, Yokoyama H, Endo T, Omi M, Tamura K, Wada N (1998) Pattern formation in dissociated limb bud mesenchyme in vitro and in vivo. Wound Rep Reg 6:398–402
Imokawa Y, Yoshizato K (1997) Expression of *Sonic hedgehog* gene in regenerating newt limb blastemas recapitulates that in developing limb buds. Proc Natl Acad Sci US 94:9159–9164

Imokawa Y, Yoshizato K (1998) Expression of *Sonic hedgehog* gene in regenerating newt limbs. Wound Rep Reg 6:366–470
Irvin BC, Tassava RA (1998) Effects of peripheral nerve implants on the regeneration of partially and fully innervated urodele forelimbs. Wound Rep Reg 6:382–387
Iten LE, Bryant SV (1973) Forelimb regeneration from different levels of amputation in the newt, *Notophthalmus viridescens*: length, rate, and stages. W Roux Archiv 173:263–282
Iten LE, Bryant SV (1975) The interaction between the blastema and stump in the establishment of the anterior-posterior and proximal-distal organization of the limb regenerate. Dev Biol 44:119–147
Jiang Y, Jahagirdar BN, Reinhardt R, Schwarts RE, Keene CD, Ortiz-Gonzalez XR, Reyes M, Lenvik T, Lund T, Blackstad M, Du J, ldrich S, Lisberg A, Low WC, Largaespada DA, Verfaille CM (2002) Pluripotency of mesenchymal stem cells derived from adult marrow. Nature 418:41–49
Johnson RL, Tabin CJ (1997) Molecular models for vertebrate limb development. Cell 90:979–990
Johnson KJ, Scadding SR (1992) Effects of tunicamycin in retinoic acid induced respecification of positional vaues in regenerating limbs of the larval axolotl, *Ambystoma mexicanum*. Dev Dyn 193:185–192
Ju B-G, Kim W-S (1994) Pattern duplication by retinoic acid treatment in the regenerating limbs of Korean salamander larvae, *Hunobius leechi* correlates well with the extent of dedifferentiation. Dev Dyn 100:253–267
Ju B-G, Kim W-S (1998) Upregulation of cathepsin D expression in the dedifferentiating salamander limb regenerate and enhancement of its expression by retinoic acid. Wound Rep Reg 6:S349-S358
Keeble S, Maden M (1986) Retinoic acid binding protein in the axolotl: distribution in mature tissues and time of appearance during limb regeneration. Dev Biol 117:435–441
Kiffmeyer WR, Tomusk EV, Mescher AL (1991) Axonal transport and release of transferring in nerves of regenerating amphibian limbs. Dev Biol 147:392–402
Kim W-S, Stocum DL (1986a) Retinoic acid modifies positional memory in the anteroposterior axis of regenerating axolotl limbs. Dev Biol 114:170–179
Kim W-S, Stocum DL (1986b) Effects of retinoic acid on regenerating normal and double half limbs of axolotls. Roux' Archiv Dev Biol 195:243–251
Kim J-H, Auerbach J, Rodriguez-Gomez JA, Velasco I, Gavin, Lumelsky N, Lee S-H, Nguyen J, Sanchez-Pernaute R, Bankiewicz K, McKay R (2002) Dopamine neurons derived from embryonic stem cells function in an animal model of Parkinson's disease. Nature 418:50–56
Korneluk RG, Liversage RA (1984) Tissue regeneration in the amputated forelimb of Xenopus laevis froglets. Canadian J Zool 62:2383–2391
Koshiba K, Kuroiwa A, Yamamoto H, Tamura K, Ide H (1998) Expression of Msx genes in regenerating and developing limbs of axolotl. J exp Zool 282:703–714
Krause DS, Theise ND, Collector MI, Henegariu, Hwang S, Gardner R, Neutzel S, Sharkis J (2001) Multi-organ, multilineage engraftment by a single bone marrow-derived stem cell. Cell 105:369–377
Kumar A, Velloso C, Imokawa Y, Brockes JP (2000) Plasticity of retrovirus-labelled myotubes in the newt regeneration blastema. Dev Biol 218:125–136

Kurabuchi S, Inoue S (1983) Denervation effects on limb regeneration in postmetamorphic *Xenopus laevis*. Dev Growth Diff 25:463–467

Laufer E, Dahn R, Orozco OE, Yeo CY, Pisenti J, Henrique D, Abbot UK, Fallon JF, Tabin C (1997) Expression of *Radical fringe* in limb-bud ectoderm regulates apical ectodermal ridge formation. Nature 386:366–373

Lagasse E, Connors, Al-Dhalimy M, Teitsma M, Dohse M, Osborne L, Wang X, Finegold M, Weissman I, Grompe M (2000) Purified hematopoietic stem cells can differentiate to hepatocytes in vivo. Nature Med 6:1229–1234

Lebowitz P, Singer M (1970) Neurotrophic control of protein synthesis in the regenerating limb of the newt *Triturus*. Nature 225:824–827

Lewandowski M, Sun X, Martin GR (2000) Fgf8 signalling from the AER is essential for normal limb development. Nat Genet 26:460–463

Lheureux E (1975) Regeneration des members irradies de *Plurodeles waltlii* Michah (urodele), influence des qualities et orientations des greffons non-irradies. W Roux Archiv 176:303–327

Litingtung Y, Dahn RD, Li Y, Fallon JF, Chiang C (2002) *Shh* and Gli3 function as a regulatory module in amniote limb patterning. Nature 418:979–983

Liversage RA (1967) Hypophysectomy and forelimb regeneration in *Ambystoma opacum* larvae. J exp Zool 165:57–70

Lo DC, Allen F, Brockes JP (1993) Reversal of muscle differentiation during urodele limb regeneration. Proc Natl Acad Sci (USA) 90:7230–7234

Logan C, Hornbruch A, Campbell I, Lumsden A (1997) The role of Engrailed in establishing the dorsoventral axis of the chick limb. Development 124:2317–2324

Loomis CA, Kimmel, Tong CX, Michaud J, Joyner A (1998) Analysis of the genetic pathway leading to formation of ectopic apical ectodermal ridges in mouse *Engrailed-1* mutant limbs. Development 125:1137–1148

Loyd RM, Tassava RA (1980) DNA synthesis and mitosis in adult newt limbs following amputation and insertion into the body cavity. J exp Zool 214:61–69

Ludolph D, Cameron JA, Stocum DL (1990) The effect of retinoic acid on positional memory in the dorsoventral axis of regenerating axolotl limbs. Dev Biol 140:41–52

Ludolph DC, Cameron JA, Stocum DL (1993a) Test of a model for the effects of retinoic acid on urodele limb regeneration. Dev Dyn 198:77–85

Ludolph DC, Cameron JA, Neff AW, Stocum DL (1993b) Cloning and tissue specific expression of the axolotl cellular retinoic binding protein. Dev Growth Diff 35:341–347

Maden M (1977) The regeneration of positional information in the amphibian limb. J Theor Biol 69:735–753

Maden M (1979) Neurotrophic and X-ray blocks in the blastemal cell cycle. J Embryol exp Morphol 50:169–173

Maden M (1981) Morphallaxis in an epimorphic system: size, growth control and pattern formation during amphibian limb regeneration. J Embryol exp Morph 65:151–167

Maden M (1982) Vitamin A and pattern formation in regenerating limbs. Nature 295:672–675

Maden M (1983) The effects of vitamin A on limb regeneration in *Rana temporaria*. Dev Biol 98:409–416

Maden M (1984) Does vitamin A act on pattern formation via the epidermis or the mesenchyme? J exp Zool 230:387–392

Maden M (1997) Retinoic acid and its receptors in limb regeneration. Sem Cell Dev Biol 8:445–453

Maden M, Wallace H (1975) The origin of limb regenerates from cartilage grafts. Acta Embryol Exp 2:77–86

Manglesdorf DJ, Umesono K, Evans RM (1994) The retinoid receptors. In: The Retinoids, 2nd ed. Sporn M, Roberts AB, Goodman DS (eds) Raven Press, New York pp 319–338

Marshak DR, Gottlieb D, Gardner RL (2001) Introduction: Stem cell biology. In: Marshak DR, Gardner RL, Gottlieb D (eds) Stem cell biology. Cold Spring Harbor Laboratory Press, Cold Spring Harbor, pp 1–16

Martin GR (1998) The roles of FGFs in the early development of vertebrate limbs. Genes Dev 12:1571–1586

Matsuda H, Yokoyama H, Endo T, Tamura K, Ide H (2001) An epidermal signal regulates *Lmx-1* expression and dorsal-ventral pattern during *Xenopus* limb regeneration. Dev Biol 229:351–362

McCormick AM, Shubeita HE, Stocum D L (1988) Cellular retinoic acid bind protein: detection and quantitation in regenerating axolotl limbs. J exp Zool 245:270–276

McDevitt, D.S., Brahma, S.K., Courtois, Y., and Jeanny, J-C (1997) Fibroblast growth factor receptors and regeneration of the eye lens. Dev Dyn. 208:220-226

McGann C, Odelberg SJ, Keating MT (2001) Mammalian myotube dedifferentiation induced by newt regeneration extract. Proc Natl Acad Sci (USA) 98:13699–13703

McGinnis W, Krumlauf R (1992) Homeobox genes and axial patterning. Cell 68:283–302

Mescher AL (1976) Effects on adult newt limb regeneration of partial and complete skin flaps over the amputation surface. J exp Zool 195:117–128

Mescher AL (1996) The cellular basis of limb regeneration in urodeles. Int J Dev Biol 40:785–796

Mescher AL, Cox CA (1988) Hyaluronate accumulation and nerve-dependent growth during regeneration of larval *Ambystoma* limbs. Differentiation 38:161–168

Mescher AL, Munaim SI (1986) Changes in the extracellular matrix and glycosaminoglycan synthesis during the imnitiation of regeneration in adult newt forelimbs. Anat Rec 214:424–431

Mescher AL, Munaim SI (1988) Transferrin and the growth-promoting effect of nerves. Int Rev Cytol 100:1–26

Mescher AL, Kiffmeyer WR (1992) Axonal release of transferrin in peripheral nerves of axolotls during regeneration. In: Taban CH, Boilly B (eds) Keys for regeneration:monographs in dev biol 23:100–109

Mescher AL, Connell E, Hsu C, Patel C, Overton B (1997) Transferrin is necessary and sufficient for the neural effect on growth in amphibian limb regeneration blastemas. Dev Growth Diff 39:677–684

Mescher AL, White GW, Brokaw JJ (2000) Apoptosis in regenerating and denervated, nonregenerating urodele forelimbs. Wound Rep Reg 8:110–116

Meyers EN, Lewendoski M., Martin GR (1998) Generation of an Fgf8 mutant allelic series using a single targeted mouse line carrying Cre and Fle recombinate recognition sites. Nature Gen 18:136–141

Michalopoulos GK, DeFrances MC (1997) Liver regeneration. Science 276:60–66
Mitashov VI (1996) Mechanisms of retina regeneration in urodeles. Int J Dev Biol 40:833–844
Mittenthal JE (1981) The rule of normal neighbors: a hypothesis for morphogenetic pattern regulation. Dev Biol 88:15–26
Miyazaki K, Uchiyawa K, Imokawa Y, Yoshizato K (1996) Cloning and characterization of of cDNAs for matrix metalloproteinases of regenerating newt limbs. Proc Natl Acad Sci USA 93:6819–6824
Mochii M, Mazaki Y, Mizuno N, Hayashi H, Eguchi G (1998) Role of Mitf in differentiation and transdifferentiation of chicken pigmented epithelial cell. Dev Biol 193:47–62
Monkmeyer J, Ludolph DC, Cameron JA, Stocum DL (1992) Retinoic acid-induced change in anteroposterior positional identity in regenerating axolotl limbs is dose-dependent. Dev Dyn 193286–294
Moon AM, Capecchi MR (2000) Fgf8 is required for outgrowth and patterning of the limbs. Nat Genet 26:455–459
Moon AM, Boulet AM, Capecchi MR (2000) Normal development in conditional mutants of Fgf4. Development 127:989–996
Morgan CJ, Pledger WJ (1992) Fibroblast proliferation. In: Cohen IK, Diegelmann RF, Lindblad WJ (eds) Wound healing: biochemical and clinical aspects. WB Saunders, Philadelphia, pp 63–76
Morzlock FV, Stocum DL (1972) Neural control of RNA synthesis in regenerating limbs of the adult newt *Triturus viridescens*. Roux's Archiv Dev Biol 171:170–180
Mullen LM, Bryant SV, Torok MA, Blumberg B, Gardiner DM (1996) Nerve dependency of regeneration: the role of *Distal-less* and FGF signaling in amphibian limb regeneration. Development 122:3487–3497
Munaim SI, Mescher AL (1986) Transferrin and the trophic effect of neural tissue on amphibian limb regeneration blastemas. Dev Biol 116:138–142
Muneoka K, Fox WF, Bryant SV (1986) Cellular contribution from dermis and cartilage to the regenerating limb blastema in axolotls. Dev Biol 116:256–260
Muneoka K, Holler-Dinsmore G, Bryant SV (1986) Intrinsic control of regenerative loss in *Xenopus laevis* limbs. J exp Zool 240:47–54
Namenwirth M (1974) The inheritance of cell differentiation during limb regeneration in the axolotl. Dev Biol 41:42–56
Nardi JB, Stocum DL (1983) Surface properties of regenerating limb cells: Evidence for gradation along the proximodistal axis. Differentiation 25:27–31
Niazi IA, Saxena S (1978) Abnormal hindlimb regeneration in tadpoles of the toad *Bufo andersoni*, exposed to vitamin A. Folia Biol (Krakow) 26:3–11
Niazi IA (1996) Background to work on retinoids and amphibian limb regeneration: Studies on anuran tadpoles—a retrospect. J Biosci 21:273–297
Niazi IA, Pescetelli MJ, Stocum DL (1985) Stage dependent effects of retinoic acid on regenerating urodele limbs. Roux's Archiv Dev Biol 194:355–363
Nieuwkoop P, Faber J (1967) Normal table of *Xenopus laevis* (Daudin): a systematical and chronological survey of the development from the fertilized egg till the end of metamorphosis, 2nd ed. North Holland Pub Co, Amsterdam
Nilsson J, von Euler AM, Dalsgaard C-J (1985) Stimulation of connective tissue growth by substance P and substance K. Nature 315:61–63

Niswander L, Tickle C, Vogel A, Booth I, Martin GR (1993) FGF-4 replaces the apical ectodermal ridge and directes outgrowth and patterning of the limb. Cell 75:579–587

Noji S, Matsuo T, Koyama E, Yamaai T, Nohno T, Matsuo N, Tanaguchi S (1990) Expression pattern of acidic and basic fibroblast growth factor genes in adult rat eyes. Biochem Biophys Res Comm 168:343–349

Oberpriller JO, Oberpriller JC (1991) Cell division in adult newt cardiomyocytes. In: Oberpriller J, Oberpriller JC, Mauro A (eds) The developmental and regenerative potential of cardiac muscle. Harwood Press, New York, pp 293–311

Odelberg SJ, Kollhof A, Keating MT (2001) Dedifferentiation of mammalian myotubes induced by msx-1. Cell 103:1099–1109

Olsen BR (1999) Bone morphogenesis and embryologic development. In: Favus MJ (ed) Primer on the metabolic bone diseases and disorders of mineral metabolism, 4th ed. Lippincott, Williams and Wilkins, Philadelphia, pp 11–14

Orkin S, Morrison SJ (2002) Stem-cell competition. 2002 Nature 418:25–27

Orts-Lorca F, Genis-Galvez JM (1960) Experimental production of retinal septa in the chick embryo: differentiation of pigment epithelium into neural retina. Acta Anat 42:31–70

O'Steen WK, Walker BE (1962) Radioautographic studies of regeneration in the common newt. III. Regeneration and repair of the intestine. Anat Rec 142:179–188

Oudkhir M, Boilly B, Lheureux E, Lasalle B (1985) Influence of denervation on the regeneration of *Pleurodele* limbs: Cytophotometric study of nuclear DNA from blastema cells. Differentiation 29:116–120

Overton J (1963) Patterns of limb regeneration in *Xenopus laevis*. J exp Zool 154:153–161

Pagan SM, Ros MA, Tabin C, Fallon JF (1996) Surgical removal of limb bud Sonic hedgehog results in posterior skeletal defects. Dev Biol 180:35–40

Park C, Hollenberg MJ (1991) Induction of retinal regeneration in vivo by growth factors. Dev Biol 148:322–333

Park C, Hollenberg MJ (1993) Growth factor-induced retinal regeneration in vivo. Int Rev Cytol 146:49–71

Park I-S, Kim W-S (1999) Modification of gelatinase activity correlates with the dedifferentiation profile of regenerating axolotl limbs. Mol Cells 9:119–126

Parr BA, McMahon AP (1995) Dorsalizing signal *Wnt-7a* required for normal polarity of D-V and A-P axes of mouse limb. Nature 374:350–353

Pecorino LT, Entwistle, Brockes JP (1996) Activation of a single retinoic acid receptor isoform mediates proximodistal respecification. Curr Biol 6563–6569

Pescitelli MJ, Stocum DL (1980) The origin of skeletal structures during intercalary regeneration of larval *Ambystoma* limbs. Dev Biol 79:255–275

Piatt J (1957) Studies on the problem of nerve pattern. III. Innervaton of the regenerated forelimb in *Amblystoma*. J exp Zool 136:229–248

Pittack C, Jones M, Reh TA (1991) Basic fibroblast growth factor induces retinal pigment epithelium to generate neural retina in vitro. Development 113:577–588

Pittack C, Grunwald GB, Reh TA (1997) Fibroblast growth factors are necessary for neural retina but not pigmented epithelium differentiation in chick embryos. Development 124:805–816

Pittenger MF, Mackay AM, Beck S, Jaiswal RK, Douglas R, Mosca JD, Moorman MA, Simonetti DW, Craig S, Marshak DR (1999) Multilineage potential of adult human mesenchymal stem cells. Science 284:143–147

Pittman RN, Buettner HM (1989) Degradation of extracellular matrix by neuronal proteases. Dev Neurosci 11:361–375

Poulin ML, Patrie KM, Botelho MJ, Tassava RA, Chiu I-M (1993) Heterogeneity in expression of fibroblast growth factor receptors during limb regeneration in newts (*Notophthalmus viridescens*). Development 119:353–361

Poulin ML and Chiu I-M. (1995) Re-programming of expression of the KGFR and *bek* variants of fibroblast growth factor receptor 2 during limb regeneration in newts (*Notopthalmus viridescens*). Dev Dyn 202:378–387

Ragsdale CW, Petkovich M, Gates PB, Chambon P, Brockes JP (1989) Identification of a novel retinoic acid receptor in regenerating tissues of the newt. Nature 341654–657

Ragsdale CW, Gates PB, Brockes JP (1992a) Identification and expression pattern of a second isoform of the newt alpha retinoic acid receptor. Nuc Acid Res 20:5851

Ragsdale CW, Gates PB, Hill D, Brockes JP (1992b) Delta retinoic acid receptor isoform $\delta 1$ is distinguished by its exceptional N-terminal sequence and abundance in the limb regeneration blastema. Mech Dev 40:99–112

Raymond PA, Hitchcock, PF (1997) Retinal regeneration: common principles but a diversity of mechanisms. Adv Neurol 72:171–184

Reyer RW (1977) The amphibian eye: development and regeneration. In: Crescitelli F (ed) Handbook of sensory physiology, Vol. VII/5. Springer-Verlag, Berlin, pp 309–390

Reyes M, Lund T, Lenvik T, Aguiar D, Koodie L, Verfaille CM (2001) Purification and ex vivo expansion of postnatal human marrow mesodermal progenitor cells. Blood 98:2615–2625

Reyes M, Dudek A, Jahagirdar B, Koodie L, Marker PH, Verfaille CM (2002) Origin of endothelial progenitors in human postnatal bone marrow. J Clin Invest 109:337–346

Richmond MJ, Pollack ED (1983) Regulation of tadpole spinal nerve growth by the regenerating limb blastema in tissue culture. J Exp Zool 225:233–242

Riddle, RD, Johnson, RL, Laufer, E, Tabin, C (1993) Sonic hedgehog mediates the polarizing activity of the ZPA. Cell 75:1401–1416

Robert B, Sassoon D, Jacq C, Gehring W, Buckingham M (1989) Hox-7, a mouse homeobox gene with a novel pattern of expression during embryogenesis. EMBO J 8:91–100

Rodriguez-Esteban C, Schwabe JW, De La Pena J, Foys B, Eshelman B, Ispizua-Belmonte JC (1997) *Radical fringe* positions the apical ectodermal ridge at the dorsoventral boundary of the vertebrate limb. Nature 386:360–366

Rodriguez-Esteban C, Schwabe JW, Pena JD, Rincon-Limas DE, Magallon J, Botas J, Belmonte JC (1998) *Lhx2*, a vertebrate homologue of apterous, regulates vertebrate limb outgrowth. Development 125:3925–3934

Rowe DA, Cairns JM, Fallon JF (1982) Spatial and temporal patterns of cerll death in limb bud mesoderm after apical ectodermal ridge removal. Dev Biol 93:83–91

Roy S, Gardiner DM, Bryant SV (2000) Vaccinia as a tool for functional analysis in regenerating limbs: ectopic expression of *Shh*. Dev Biol 218:199–205

Saunders JW Jr (1948) The proximo-distal sequence of the origin of the parts of the chick wing and the role of the ectoderm. J exp Zool 108:363–403
Scadding SR, Maden M (1994) Retinoic acid gradients during limb rergeneration. Dev Biol 162:608-617
Schmidt AJ (1966) The Molecular Basis of Regeneration: Enzymes. Illinois Monographs Med Sci 6 (4), University of Illinois Press, Urbana
Schotte OE, Butler EG (1944) Phases in regeneration of the urodele limb and their dependence on the nervous system. J exp Zool 97:95–121
Shimizu-Nishikawa K, Tazawa I, Uchiyama K, Yoshizato K (1999) Expression of helix-loop-helix type negative regulators of differentiation during limb regeneration in urodeles and anurans. Dev Growth Diff 41:731–743
Shimizu-Nishikawa KS, Tsuji S, Yoshizato K (2001) Identification and characterization of newt rad (ras associated with diabetes), a gene specifically expressed in regenerating limb muscle. Dev Dyn 220:74–86
Shubin NH, Alberch P (1986) A morphogenetic approach to the origin and basic organization of the tetrapod limb. Evol Biol 20:318–390
Schwartz RE, Reyes M, Koodie L, Jiang Y, Blackstad M, Lund T, Lenvik T, Johnson S, Hu W-S, Verfaille M (2002) Multipotent adult progenitor cells from bone marrow d8ifferentiate into functional hepatocyte-like cells. J Clin Invest 109:1291–1302
Sessions SK, Bryant V (1988) Evidence that regenerative ability is an intrinsic property of limb cells in *Xenopus*. J exp Zool 247:39–44
Simon H-G, Tabin CJ. 1993. Analysis of *Hox- 4.5* and *Hox-3.6* expression during newt limb regeneration: differential regulation of paralogous *Hox* genes suggests different roles for members of different *Hox* clusters. Development 111:1397–1407
Simon H-G, Nelson C, Goff D, Laufer E, Morgan BA, Tabin C (1995) Differential expression of myogenic regulatory genes and Msx-1 during dedifferentiation and redifferentiation of regenerating amphibian limbs. Dev Dyn 202:1–12
Simon H-G, Kittappa R, Han PA, Tsilfildis C, Liversage RA, Oppenheimer S (1997) A novel family of T-box genes in urodele amphibian limb development and regeneration: candidate genes involved in vertebrate forelimb/hindlimb patterning. Development 124:1355–1366
Singer M (1952) The influence of the nerve in regeneraton of the amphibian extremity. Quart Rev Biol 27:169–200
Singer M (1965) A theory of the trophic nervous control of amphibian limb regeneration, including a re-evaluation of quantitative nerve requirements. In: Kiortsis V, Trampusch HAL (eds) Regeneration in animals and related problems. North Holland, Amsterdam, pp. 20–32
Singer M (1978) On the nature of the neurotrophic phenomenon in urodele limb regeneration. Am Zool 18:829–841
Singer M, Craven L (1948) The growth and morphogenesis of the regenerating forelimb of adult *Triturus* following denervation at various stages of development. J Exp Zool 108:279–308
Singer M, Nordlander RH, Egar P (1979) Axonal guidance during embryogenesis and regeneration in the spinal cord of newt: the blueprint hypothesis of neuronal pathway patterning. J Comp Neurol 185:1–22
Slack JMW, Savage S (1978) Regeneration of reduplicated limbs in contravention of the complete circle rule. Nature 271:760–761

Smith AR, Lewis JH, Crawly A, Wolpert L (1974) A quantitative study of blastemal growth and bone regression during limb regeneration in *Triturus cristatus*. J Embryol exp Morph 32:375–390

Smith GN, Toole BP, Gross J (1975) Hyaluronidase activity and glycosaminoglycan synthesis in the amputated newt limb: Comparison of denervated nonregenerating limbs with regenerates. Dev Biol 43:221–232

Smith MJ, Globus M (1989) Multiple interactions in juxtaposed monolayer of amphibian neuronal, epidermal, and mesodermal limb blastema cells. In Vitro Cell Dev Dev Biol 25:849–856

Song K, Wang Y, Sassoon D (1992) Expression of *Hox-7.1* in myoblasts inhibits terminal differentiation of and induces cell transformation. Nature 360:477–481

Sperry RW (1944) Optic nerve regeneration with return of vision in amphibians. J Neurophysiol 7:57–69

Stark RJ, Searls RL (1973) A description of chick wing development and a model of limb morphogenesis. Dev Biol 33:138–153

Stark DR, Gates PB, Brockes JP, Ferretti P (1998) *Hedgehog* family member is expressed throughout regenerating and developing limbs. Dev Dyn 212:352–363

Steen TP (1968) Stability of chondrocyte differentiation and contribution of muscle to cartilage during limb regeneration in the axolotl (*Siredon mexicanum*). J exp Zool 167:49–78

Steen TP (1973) The role of muscle cells in *Xenopus* limb regeneration. Am Zool 13:1349–1350

Steen TP, Thornton CS (1963) Tissue interaction in amputated aneurogenic limbs of *Ambystoma* larvae. J exp Zool 154:207–221

Stensaas LJ (1983) Regeneration in the spinal cord of the newt *Notophthalmus (Triturus) pyrrhogaster*. In: Kao CC, Bunge RP, Reier PJ (eds). Spinal cord reconstruction, Raven Press, New York, pp 121–149

Steinberg MS (1978) Cell-cell recognition in multicellular assembly: levels of specificity. In: Curtis ASG (ed). Cell-cell recognition. Cambridge University Press, Cambridge, pp 25–49

Stocum DL (1968a) The urodele limb regeneration blastema: A self-organizing system. I. Differentiation in vitro. Dev Biol 18:441–456

Stocum D (1968b) The urodele limb regeneration blastema: self-organizing system. II. Morphgenesis and differentiation of autografted whole and fractionall blastemas. Dev Biol 18:457–480

Stocum DL (1975) Regulation after proximal or distal transposition of limb regeneration blastemas and determination of the proximal boundary of the regenerate. Dev Biol 45:112–136

Stocum DL (1978a) Organization of the morphogenetic field in regenerating amphibian limbs. Am Zool 18:883–896

Stocum DL (1978b) Regeneration of symmetrical hindlimbs in larval salamanders. Science 200:790–793

Stocum DL (1979) Stages of forelimb regeneration in *Ambystoma maculatum*. J exp Zool 209:395–416

Stocum DL (1980a) Autonomous development of reciprocally exchanged regeneration blastemas of normal forelimbs and symmetrical hindlimbs. J exp Zool 212:361–371

Stocum DL (1980b Intercalary regeneration of symmetrical thighs in the axolotl, *Ambystoma mexicanum*. Dev Biol 79:276–295
Stocum DL (1981) Distal transformation in regenerating double anterior axolotl limbs. J Embryol exp Morph 65:3–18
Stocum DL (1982) Determination of axial polarity in the urodele regeneration blastema. J Embryol Exp Morph 71:193–214
Stocum DL (1983) Amphibian limb regeneration: distal transformation. In: Fallon JF, Caplan AI (eds). Limb development and regeneration, Part A. Alan R. Liss, Inc, New York, pp. 467–476
Stocum DL (1984) The urodele limb regeneration blastema: Determination and organization of the morphogenetic field. Differentiation 27:13–28
Stocum DL (1991) Limb regeneration: a call to arms (and legs). Cell 67:5–8
Stocum DL (1995) Wound Repair, Regeneration, and Artificial Tissues. RG Landes Co, Austin, 230 pp
Stocum DL (1996) A conceptual framework for analyzing axial patterning in regenerating urodele limbs. Int J Dev Biol 40:773–784
Stocum DL (2000) Transplantation and culture techniques for the analysis of urodele limb regeneration. In: Tuan RS, Lo CW (eds). Developmental biology protocols, Vol I. Humana Press, Towota, pp 67–74
Stocum DL (2001) Stem cells in regenerative biology and medicine. Wound Rep Reg 9:429–442
Stocum DL, Dearlove GE (1972) Epidermal-mesodermal interaction during morphogenesis of the limb regeneration blastema in larval salamanders. J exp Zool 181:49–62
Stocum DL, Melton DA (1977) Self-organizational capacity of distally transplanted limb regeneration blastemas in larval salamanders. J exp Zool 201:451–472
Stocum DL, Crawford K (1987) Use of retinoids to analyze the cellular basis of positional memory in regenerating axolotl limbs. Biochem Cell Biol 65:750–761
Stocum DL, Fallon JF (1982) Control of pattern formation in urodele limb ontogeny: A review and a hypothesis. J Embryol Exp Morphol 69:7–36
Stocum DL, Maden M (1990) Regenerating limbs. Methods in Enzymol 190:189–201
Stocum DL, Mitashov VI (1990) Pattern regulation in regenerating limbs. Ontogenez 27:5–31. In Russian, English translation by Plenum Publishing Corp
Stroeva OG, Mitashov VI (1983) Retinal pigment epithelium: proliferation and differentiation during development and regeneration. Int Rev Cytol 38:221–293
Summerbell D, Lewis JR (1975) Time, place and positional vaue in the chick limb bud. J Embryol exp Morph 33:621–643
Summerbell D, Lewis JR, Wolpert L (1973) Positional information in chick limb morphogenesis. Nature 244:228–230
Sun X, Lewandowski M, Meyers EN, Liu YH, Maxson RE, Martin GR. (2000) Conditional inactivation of FGF4 reveals complexity of signaling during limb bud development. Nat Genet 25:83–86
Sun X, Mariani FW, Martin GR (2002) Functions of FGF signaling from the apical ectodermal ridge in limb development. Nature 418:501–508
Sussman HH (1989) Iron and tumor cell growth. In: deSousa M, Brock JH (eds). Iron in immunity, cancer and inflammation. John Wiley and Sons, New York, pp 261–282

Swalla BJ, Solursh M (1984) Inhibition of limb chondrogenesis by fibronectin. Differentiation 26:42–48
Tabin C (1989) Isolation of potential limb identity genes. Development 105:813–820
Tamura K, Yokouchi Y, Kuroiwa A, Ide H (1997) Retinoic acid changes the proximodistal developmental competence and affinity of distal cells in the developing chick limb bud. Dev Biol 188:224–234
Tanaka EM, Gann F, Gates PB, Brockes JP (1997) Newt myotubes re-enter the cell cycle by phosphorylation of the retinoblastoma protein. J Cell Biol 136:155–165
Tanaka EM, Dreschel N, Brockes JP (1999) Thrombin regulates S phase re-entry by cultured newt myoblasts. Curr Biol 9:792–799
Tank PW (1978) The occurrence of supernumerary limbs following blastemal transplantation in the regenerating forelimb of the axolotl, *Ambystoma mexicanum*. Dev Biol 62:143–161
Tank PW, Carlson BM, Connelly TG (1976) A staging system for forelimb regeneration in the axolotl, *Ambystoma maculatum*. J Morph 150:117–128
Tassava RA, Acton RD (1989) Distribution of a wound epithelium antigen in embryonic tissues of newts and salamanders. Ohio J Sci 89 (1):12–15
Tassava RA, Garling DJ (1979) Regenerative responses in larval axolotl limbs with skin grafts over the amputation surface. J exp Zool 208:97–110
Tassava RA, Mescher AL (1975) The roles of injury, nerves and the wound epidermis during the initiation of amphibian limb regeneration. Differentiation 4:23–24
Tassava RA, Chlaponski FA, Thornton CS (1968) Limb regeneration in *Ambystoma* larvae during and after treatment with adult pituitary hormone. J exp Zool 167:157–167
Teitelbaum SL (2000) Bone resorption by osteoclasts. Science 289:1504–1508
Terada N, Hamazaki T, Oka M, Hoki M, Mastalerz DM, Nakano Y, Meyer EM, Morel L, Petersen BE, Scott EW (2002) Bone marrow cells adopt the phenotype of other cells by spontaneous cell fusion. Nature 416:542–545
Thoms SD, Stocum DL (1984) Retinoic acid-induced pattern duplication in regenerating urodele limbs. Dev Biol 103:319–328
Thornton CS (1938) The histogenesis of muscle in the regenerating forelimb of larval *Ambystoma punctatum*. J Morph 62:17–46
Thornton CS (1960) Influence of an eccentric epidermal cap on limb regeneration in *Amblystoma* larvae. Dev Biol 2:551–569
Thornton CS (1962) Eccentric blastema formation in aneurogenic limbs of*Ablystoma* following epidermal cap deviation. Dev Biol 5:328–343
Thornton CS (1968) Amphibian limb regeneration. Adv Morph 7:205–249
Thornton CS, Thornton MT (1965) The regeneration of accessory limb parts following epidermal cap transplantation in urodeles. Experientia 21:1–6
Tomlinson BL, Tassava RL (1987) Dorsal root ganglia stimulate regeneration of denervated urodele forelimbs: timing of graft implantation with respect to denervation. Development 99:173–186
Tonge DA, Leclere PG (2000) Directed axonal growth towards axolotl limb blastemas in vitro. Neurosci 100:201–211
Toole BP, Jackson GJ, Gross J (1972) Hyaluronate in morphogenesis: inhibition of chondrogenesis in vitro. Proc Natl Acad Sci USA 69:1384–1386

Torok MA, Gardiner M, Shubin N, Bryant SV (1998) Expression of *HoxD* genes in developing and regenerating axolotl limbs. Dev Biol 200:225–233

Torok MA, Gardiner DM, Izpisua-Belmonte J-C, Bryant SV (1999) *Sonic hedgehog* (*shh*) expression in developing and regenerating axolotl limbs. J exp Zool 284:197–206

Tschumi PA (1957) The growth of hindlimb bud of *Xenopus laevis* and its dependence upon the epidermis. J Anat 91149-173

Tsonis PA, Trombley MT, Rowland T, Chandraratna RAS, Del-Rio Tsonis K (2000) Role of retinoic acid in lens regeneration. Dev Dyn 219:588–593

Van Stone JM (1955) The relationship between innervation and regenerative capacity in hind limbs of *Rana sylvatica*. J Morph 97:345–391

Velloso CP, Kumar A, Tanaka EM, Brockes JP (2000) Generation of mononucleate cells from post-mitotic myotubes proceeds in the absence of cell cycle progression. Differentiation 66:239–246

Velloso CP, Simon A, Brockes JP (2001) Mammalian postmitotic nuclei reenter the cell cycle after serum stimulation in newt/mouse hybrid myotubes. Curr Biol 11:855–858

Viviano CM, Horton CE, Maden M, Brockes JP (1995) Synthesis and release of 9-*cis* retinoic acid by the urodele wound epidermis. Development 121:3753–3762

Vogt TF, Duboule D (1999) Antagonists go out on a limb. Cell 99:563–566

Vorontsova MA, Liosner LD (1960) Asexual propagation and regeneration. Permagon Press, New York pp 143–159

Vortkamp A (2001) Interaction of growth factors regulating chondrocyte differentiation in the developing embryo. Osteoarth Cartilage 9 (Suppl A):S109-S117

Wada N, Uchiyama K, Ide H (1993) Cell sorting out and chondrogenic aggregate formation in limb bud recombinant and in culture. Dev Growth Diff 35:421–430

Wada N, Ide H (1994) Sorting out of limb bud cells in monolayer culture. Int J Dev Biol 38:351–356

Wada N, Kimura I, Tanaka H, Ide H, Nohno T (1998) Glyccosylphosphatidylinositol-anchored cell surface proteins regulate position-specific cell affinity in the limb bud. Dev Biol 202:244–252

Wang L, Marchionni MA, Tassava RA (2000) Cloning and neuronal expression of a type III newt neuregulin and rescue of denervated nerve-dependent newt limb blastemas by rhGGF2. J Neurobiol 43:150–158

Weissman IL (2000a) Translating stem and progenitor cell biology to the clinic: barriers and opportunities. Science: 287:1442–1446

Weissman IL (2000b) Stem cells: units of development, units of regeneration, and units in evolution. Cell 100:157–168

Wolfe D, Nye HLD, Cameron J (2000) Extent of ossification at the amputation plane is correlated with the decline of blastema formation and regeneration in *Xenopus laevis* hindlmbs. Dev Dyn 218:681–697

Woloshin P, Song K, Degnin A, Killary DJ, Goldhamer DJ, Sassoon D, Thayer MJ (1995) MSX1 inhibits MyoD expression in fibroblast X 10T1/2 cell hybrids. Cell 82:611–620

Yajima H, Yonei-Tamura S, Watanabe N, Tamura K, Ide H (1999) Role of N-cadherin in the sorting-out of mesenchymal cels and in the positional identity along the proximodistal axis of the chick limb bud. Dev Dyn 216:274–284

Yang EV, Gardiner DM, Bryant SV (1999) Expression of *Mmp-9* and related matrix metalloproteinase genes during axolotl limb regeneration. Dev Dyn 216:2–9

Yannas IV (2001) Tissue and organ regeneration in adults. Springer-Verlag New York, Inc, New York, pp 138–185

Ying Q-L, Nichols J, Evans EP, Smith AG (2002) Changing potency by spontaneous fusion. Nature 416:545–548

Yntema CL (1959a) Regeneration of sparsely innervated and aneurogenic forelimbs of *Ambystoma* larvae. J exp Zool 140:101–123

Yntema CL (1959b) Blastema formation in sparsely innervated and aneurogenic forelimbs in *Amblystoma* larvae. J exp Zool 142:423–440

Yokoyama H, Yonei-Tamura S, Endo T, Izpisua-Belmonte JC, Tamura K, Ide H (2000) Mesenchyme with *fgf10* expression is responsible for regenerative capacity in *Xenopus* limb buds. Dev Biol 219:18–29

Yokoyama H, Ide H, Tamura K (2001) FGF-10 stimulates limb regeneration ability in *Xenopus laevis*. Dev Biol 233:72–79

Young HE, Bailey CF, Dalley BK (1983) Gross morphological analysis of limb regeneration in postmetamorphic adult *Ambystoma*. Anat Rec 206:295–306

Young HE, Dalley B, Markwald RR (1989) Effect of selected denervations on glycoconjugate composition and tissue morphology during the initiation phase of limb regeneration in adult *Ambystoma*. Anat Rec 223:230

Zardoya R, Abouheif A, Meyer A (1996) Evolution and orthology of hedgehog genes. Trends Genet 12:496–497

Zenjari C, Boilly, Hondermarck H, Boilly-Marer Y (1997) Nerve-blastema interactions induce fibroblast growth factor-1 release during limb regeneration in *Pleurodeles waltl*. Dev Growth Diff 39:15–22

Existence of Reserve Quiescent Stem Cells in Adults, From Amphibians to Humans

H. E. Young
Division of Basic Medical Sciences and Department of Pediatrics,
Mercer University School of Medicine, 1550 College Street, Macon, GA 31207, USA
E-mail: young_he@mercer.edu

1 Theories for Tissue Restoration 72
2 Limb Regeneration in Adult Terrestrial Salamanders 73
3 Existence of Quiescent Stem Cells in Higher Order Animals 81
4 Phylogenetic Distribution 85
5 Age of Donor 87
6 Location of Pluripotent Stem Cells 87
7 Effects of Bioactive Factors 88
8 Unique Cell Surface Antigens 88
9 Pluripotent Stem Cells Residing in Tissues 89
10 Pluripotent Mesenchymal Stem Cells 89
11 Pluripotent Epiblastic-Like Stem Cells 91
12 Capability for Extended Self-Renewal 95
13 Paradigm Shift 97
14 Activation of Reserve Quiescent Stem Cells 103
References 104

Abstract Several theories have been proposed to explain the phenomenon of tissue restoration in amphibians and higher order animals. These theories include dedifferentiation of damaged tissues, transdifferentiation of lineage-committed stem cells, and activation of quiescent stem cells. Young and colleagues demonstrated that connective tissues throughout the body contain multiple populations of quiescent lineage-committed progenitor stem cells and lineage-uncommitted pluripotent stem cells. Subsequent cloning and cell sorting studies identified quiescent lineage-uncommitted pluripotent mesenchymal stem cells, capable of forming any mesodermal cell type, and pluripotent epiblastic-like stem cells, capable of forming any somatic cell type. Based on their studies, they propose at least 11 categories of quiescent reserve stem cells resident within postnatal animals, including humans. These categories are pluripotent epiblastic-like stem cells, pluripotent ectodermal stem cells, pluripotent epidermal stem cells, pluripotent neuronal stem cells, pluripotent neural crest stem cells, pluripotent mesenchymal (mesodermal) stem cells, pluripotent endodermal stem cells, multipotent progenitor stem cells, tripotent progenitor stem cells, bipotent progenitor stem cells, and unipotent progenitor stem cells. Thus, activation of quiescent reserve stem cells, i.e., lineage-committed progenitor stem cells and lineage-uncommitted pluripotent stem cells, resident within the connective tissues could provide for the continual maintenance and repair of the postnatal organism after birth

1
Theories for Tissue Restoration

Tissue restoration is the 'holy grail' of tissue engineering. One classic example of tissue restoration is limb regeneration in amphibians. After transection of the limb, an epidermal cap/ridge forms by re-epithelialization of the transected portion of the limb. Concomitant with re-epithelialization is the formation of an underlying mass of undifferentiated cells, termed the blastema. Once formed, signals emanating from the epidermal cap/ridge interact with the blastemal cells to initiate their proliferation, subsequent outgrowth, and restoration of the limb structures (Fig. 1, parts 1–14).

Several theories have been proposed to explain the phenomenon of tissue restoration in amphibians and higher order animals. One such theory is dedifferentiation. Toole and Gross (1971) reported that blastema formation in the juvenile aquatic salamander occurred by the process of dedifferentiation of the transected tissues. The dedifferentiation theory is predicated on the belief that stem cells either do not exist within the tissues or do not function in the process of tissue restoration. Rather, differentiated cells damaged during the initial trauma revert to a more primitive undifferentiated state, forming the blastema, prior to their proliferation and subsequent re-differentiation to restore the missing tissues (for a review see Stocum 1998).

The theory of transdifferentiation offers an alternate hypothesis to explain the phenomenon of tissue restoration in amphibians and higher order animals. This theory is predicated on the belief that only tissue-specific lineage-committed stem cells exist within a postnatal individual and that their normal function is to maintain and/or replace their respective tissue type (for a review see Donovan and Gearhart 2001). The process of transdifferentiation occurs during tissue restoration when tissue-specific lineage-committed stem cells derived from one organ are reprogrammed to form tissues of another organ. For example, stem cells derived from the bone marrow have been reported to form neurons and neural supportive tissues (Eglitis and Mezey 1997; Woodbury et al. 2000), as well as hepatic oval cells (Petersen et al. 1999). Stem cells derived from neuronal tissues have been reported to form blood elements (Bjornson et al. 1999; Vescovi et al. 2001), and stem cells derived from skeletal muscle have been reported to form blood (Jackson et al. 1999). These investigators propose that the tissue-specific lineage-committed stem cells from one organ are reprogrammed or transdifferentiated to form cells and tissues of another organ.

Young and colleagues have proposed that activation of quiescent stem cells underlies the phenomenon of tissue restoration (Young 1983, 2000; Young et al. 1983a–c, 1985, 1989a,b, 1991, 1992a,b, 1993, 1995, 1998a,b, 1999, 2001a,b, 2003a–c; Lucas et al. 1993, 1995, 1996a,b; Pate et al. 1993; Rogers et al. 1995; Dixon et al. 1996; Warejcka et al. 1996; Young and Black 2003). This hypothesis postulates that lineage-committed progenitor stem cells and lineage-uncommitted pluripotent stem cells are present as reserve populations of stem cells throughout the body. These cells are involved in the maintenance and replacement of tissues of the body. During tissue restoration these quiescent stem cells become activated, proliferate, and differentiate into the missing tissues.

2
Limb Regeneration in Adult Terrestrial Salamanders

It has been well established that aquatic-phase urodeles, i.e., the juvenile and the adult newt, the juvenile and the adult axolotl, and the juvenile larval salamander, have the capacity to completely regenerate an appendage after traumatic or surgical amputation (Thornton 1968; Singer 1978; Tank and Holder 1981). In these species complete restoration of the missing tissues was observed during a relatively short period of

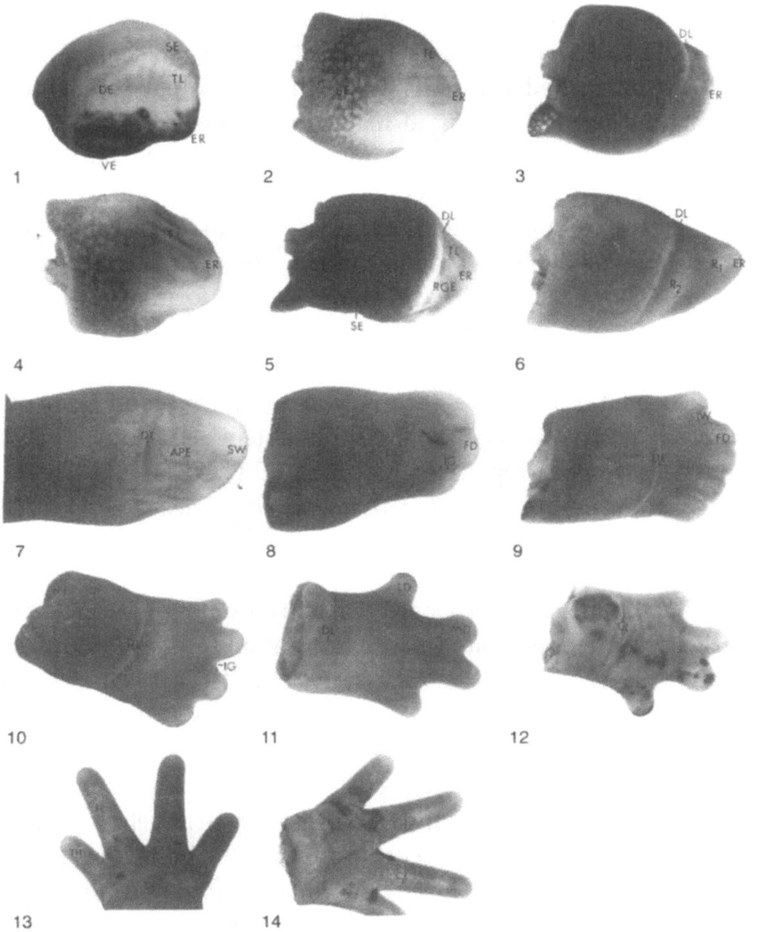

Fig. 1 (*1*) Wound healing stage, dorsal oblique view, shows a slight swelling in the regenerate epidermis at the distal tip and the first evidence for the bud outgrowth, *Ambystoma annulatum*, ×100. (*2*) Early bud stage, ventral view, shows the rounded outgrowth of the blastema with the epidermal ridge at its distal tip, *A. annulatum*, ×70. (*3*) Middle bud stage, dorsal view, shows a demarcation line separating the stump tissues from a symmetrical cone-shaped blastemal outgrowth, *A. annulatum*, ×94. (*4*) Middle bud stage, ventral view of the same bud as in (*3*). Tension lines are visible from the epidermal ridge to the stump tissues, *A. annulatum*, ×90. (*5*) Late bud stage, dorsal view of a flattened cone-shaped outgrowth. The demarcation lines continues to separate the blastemal outgrowth from the remaining portion of the limb, *A. annulatum*, ×60. (*6*) Early palette stage, dorsal view shows a paddle-shaped

time: 40 days for the newt (Iten and Bryant 1973), '30+' days for the axolotl (Farber 1959; Tank et al. 1976), and 44 days for the juvenile larval salamander (Stocum 1979). In salamanders, once metamorphosis occurs and the juvenile larval aquatic form transforms into an adult terrestrial form, complete absence of regeneration or only heteromorphic regenerates were reported (Scadding 1977). Thus it was generally accepted that the adult terrestrial form lost its intrinsic ability to regenerate any ap-

outgrowth which can be segregated into two regions: region 1 comprises the area of the future digits; region 2 comprises the area of the future 'hand', *A. annulatum*, ×70. (*7*) Middle palette stage, dorsal view of an elongated palette-shaped outgrowth. The apical palette epidermis covers the outgrowth and a demarcation line separates the stump tissues from regenerate outgrowth tissues. Along the distal ridge of the outgrowth, slight swellings begin to replace the epidermal ridge, *A. annulatum*, ×70. (*8*) Late palette stage, dorsal view of elongated palette-shaped outgrowth. A demarcation line separates proximal stump tissues from the distal regenerate tissues. Distinct swellings are visible along the distal border and are indicative of forming digits, *A. annulatum*, ×90. (*9*) Early digit stage, dorsal view of the regenerate structure that has the distinct appearance of a forming 'hand' containing digits. The forming digits are beginning to separate from each other, *A. annulatum*, ×60. (*10*) Middle digit stage, dorsal view of a definitive 'hand'-like regenerate structure consisting of separated digital regions, *A. annulatum*, ×70. (*11*) Late digit stage, a dorsal view showing continued separation of the digits, *A. annulatum*, ×70. (*12*) Late digit stage, a ventral view of the same regenerate as seen in (*11*). Patches of chromatophores (melanocytes) are appearing along this surface. Transverse crease lines overlie future joint areas, such as those for the interphalangeal joints, *A. annulatum*, ×70. (*13*) Complete limb regenerate stage, a dorsal view showing all digits separated from each other with the regenerate forelimb indistinguishable from either the original or sham-operated control limbs. Within the regenerated limb the demarcation line remains and marks the spot of the original amputation site, *A. annulatum*, ×60. (*14*) Complete regenerate stage, a ventral view of (*13*). At the beginning of this stage, patches of chromatophores appear along the ventral surface, but by the end of this stage the pigmentation pattern assumes that of its respective species, i.e., annular rigns for *A. annulatum*, spots for *A. maculatum*, a mottled appearance for *A. texanum*, and stripes for *A. tigranum*, *A. annulatum*, ×60. *APE*, apical pallette epidermis; *C*, chromatophores; *CL*, crease lines; *DE*, dorsal regenerative epidermis; *DL*, demarcation line; *ER*, epidermal ridge; *FD*, forming digits; *FL*, finger-like digits; *IG*, interdigital grooves; *IW*, interdigital wedge; *LD*, lateral digits; *MD*, middle digits; *R1*, distal one-third of outgrowth, region 1; *R2*, proximal two-thirds of outgrowth, region 2; *RGE*, regenerate epidermis; *SE*, stump epidermis; *SW*, swellings; *TH*, thumb-like digit; *TL*, tension lines; *VE*, ventral regenerate epidermis. (From Young et al. 1983b Anatomical Record 206:295–306, Reprinted with permission from Alan R. Liss, Inc.)

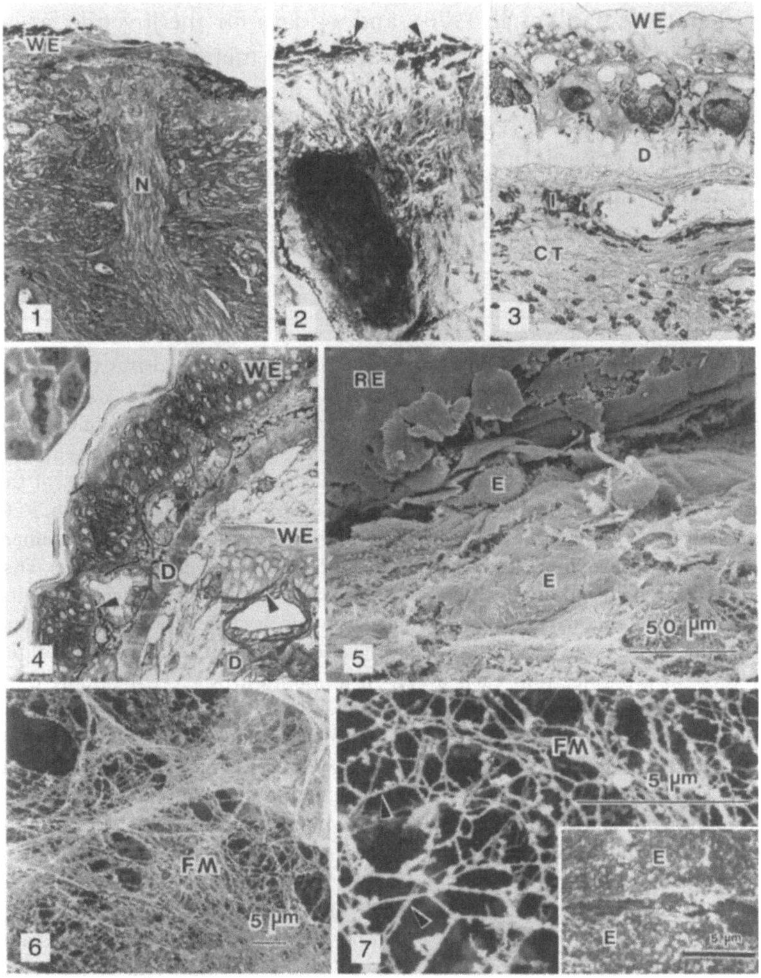

Fig. 2 (*1*) Wound epidermis surrounding amputation site 18–20 h after amputation of normal limb. Neural tissue along basal border of wound epidermis, ×300. (*2*) Limb tissues transected by amputation 18–20 h after amputation. Arrowheads denote degenerating tissues along amputation surface, ×450. (*3*) Postaxial border of stump tissue area just proximal to amputation site, 18–20 h after amputation. Stump epidermis adjacent to amputation site demonstrates intracellular staining for hyaluronic acid and glycoproteins. Dermis consists of dense fibrous connective tissue. Inflammatory cells (macrophages and neutrophils) located within and around blood vessel periphery and interspersed within connective tissues, ×900. (*4*) Postaxial stump tissue area proximal to amputation site, 24–26 h after amputation. Stump epidermis adjacent to amputation sites shows intracellular staining for hyaluronic acid

pendage. The mechanism of action for loss of this regenerative ability in the adult urodele salamander was postulated, but unknown at that time (Singer 1951; Pritchette and Dent 1972).

Young et al. (1983a) reported that adverse laboratory conditions accounted for the apparent loss of regenerative ability in adult terrestrial salamanders. When the laboratory environment was adjusted for terrestrial conditions, adult salamanders, i.e., *Ambystoma annulatum*, *A. maculatum*, *A. texanum*, and *A. tigranum*, completely regenerated their limbs (Fig. 1, parts 1–14) within 155–370 days postamputation (Young 1983; Young et al. 1983b,c, 1985). Additional studies (Young et al. 1989a, 1989b) noted that the transected tissues of the limb were completely removed by a debridement/inflammatory response by neutrophils and macrophages (Fig. 2, parts 2 and 3). The cells forming the blastema already existed in the adult salamanders as quiescent cells. They were located within the more proximal supporting connective tissues of the skeletal muscle, dermis, fat, loose fibrous connective tissues, cartilage,

◄

and glycoproteins. Epidermal basement membrane (*arrowhead*) and dermis display heavy staining for glycoproteins, ×600. Lower right insert is preaxial border of limb stump. Stump epidermis adjacent to amputation site demonstrates intracellular staining for only hyaluronic acid. Epidermal basement membrane and dermis show staining for glycoproteins, ×600. Upper left insert demonstrates presence of mitotic figure within stump epidermis surrounding amputation site, ×1,800. (*5*) En face view of wound surface surrounding cuff of regenerate epidermis 60 h after amputation. Flattened epithelial-like cells covering wound surface, scanning electron microscopy. (*6*) En face view of amputation surface 72 h after amputation. Fine filamentous material located on outer surface of flattened epithelial-like cells covering the wound surface of a *Streptomyces hyaluronidase* buffer control limb, scanning electron microscopy. (*7*) Higher magnification of filamentous material covering amputation site 72 h after amputation. *Arrowheads* demarcate fine filamentous material, SEM. Lower right insert reveals similar area after digestion with *Streptomyces hyaluronidase*. This insert demonstrates the sensitivity (complete loss) of the wound surface filamentous material (hyaluronic acid) after incubation with the enzyme. External surface of two flattened epithelial-like cells covering the wound surface are visible. *AEC*, apical epidermal cap; *B*, basal layers of regenerate epidermis; *C*, cartilage; *CB*, core blastema; *CT*, loose connective tissues; *D*, dermis; *E*, epithelial-like cells; *F*, free cell population; *FM*, fine filamentous material; *G/GM*, granulated wound surface matrix; *I*, inflammatory cells (macrophages and neutrophils); *N*, nerve; *RE*, regenerate epidermis; *SREB*, subregenerate epidermal blastema; *ST*, stump tissues; *TP*, trailing cells; *WE*, wound epidermis. (From Young et al. 1989a Anatomical Record 223:231–241, Reprinted with permission from Alan R. Liss, Inc.)

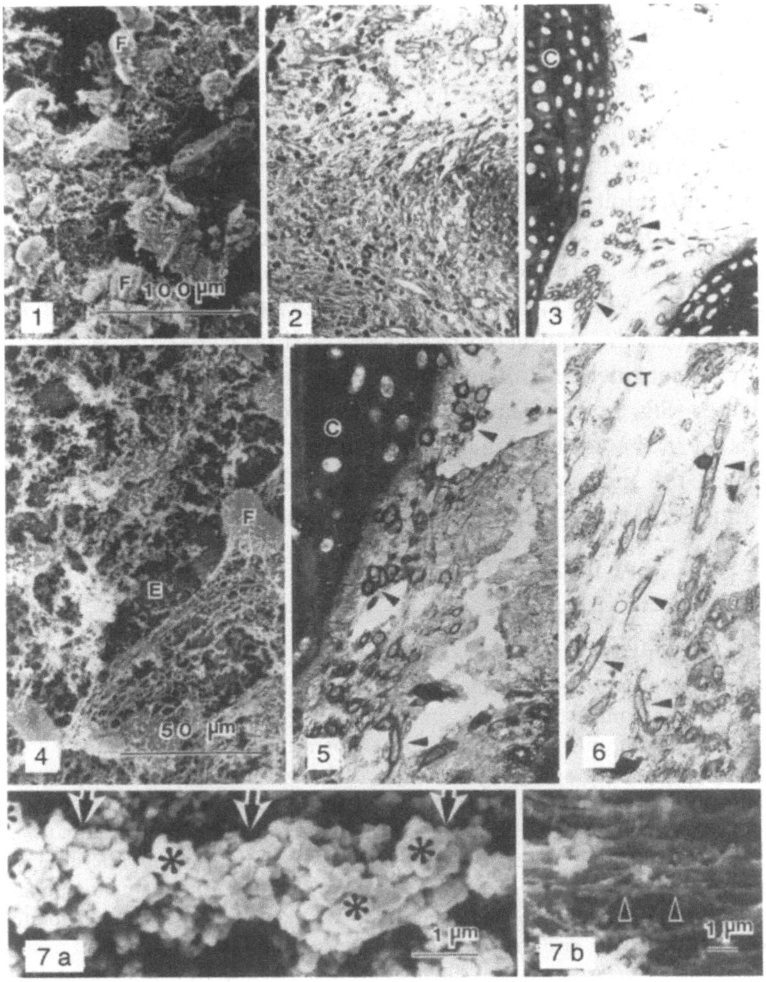

Fig. 3 (*1*) Free cell population located within filamentous wound surface matrix near regenerate epidermal cuff 84 h after amputation, SEM. (*2*) Area of stump tissues proximal to the amputation site 5 days after amputation. Extracellular hyaluronic acid interspersed with blastemal cells and connective tissue cells surrounding vessels, ×500. (*3*) Proximal area of stump tissues between two cartilage masses 5 days after amputation. *Arrowheads* demonstrate cells with intracellular heparan sulfate glycosaminoglycans, pericellular heparan sulfate glycosaminoglycans and glycoproteins, and extracellular hyaluronic acid. This figure reveals putative (mesenchymal-like) blastemal cells associated with the perichondrial sheath covering the cartilage and within the connective tissues between the cartilage masses, ×600. (*4*) En face view of wound surface matrix 6 days after amputation. There is a continued associa-

and bone of the limb (Figs. 3, parts 3, 5, and 6). Before amputation, no differences between the quiescent cells and the remaining undifferentiated cells of the connective tissues were detected. Connective tissue populations assessed were the endomysium, perimysium, and epimysium of skeletal muscle; perichondrium; periosteum; dermis; the endoneurium, perineurium, and epineurium of nerves; connective tissues associated with fat; and general loose fibrous connective tissues. Immediately after amputation of the limb, a unique population of cells would begin synthesizing heparan sulfate proteoglycan. These same cells would then show a transient shift from intracellular to pericellular heparan sulfate proteoglycan staining. Glycoproteins emanating from the forming epidermal cap (Fig. 4, part 2) would bind to pericellular heparan sulfate proteoglycans. This unique population of cells with cell surface-bound heparan

tion of the free cell population with the filamentous (hyaluronic acid) wound surface matrix along the external surface of the epithelial-like cells. There is also co-localization of 50-nm granules with the free cell population within the filamentous wound surface matrix, SEM. (*5*) Area of the stump tissues proximal to the amputation site 6 days after amputation. *Arrowheads* denote cells with intracellular heparan sulfate glycosaminoglycans, pericellular heparan sulfate glycosaminoglycans and glycoproteins, and extracellular hyaluronic acid. These cells are the putative (mesenchymal-like) blastemal cells that are associated with the perichondrium and associated connective tissues adjacent to the cartilage mass, ×1200. (*6*) Area of the connective tissues located between skeletal muscle and adjacent cartilage 6 days after amputation. This area is within the tissues proximal to the amputation site. Putative (mesenchymal-like) blastemal cells containing intracellular heparan sulfate glycosaminoglycans, pericellular heparan sulfate glycosaminoglycans and glycoproteins, and extracellular hyaluronic acid (*arrowheads*) interspersed within the connective tissue, ×1,200. (*7a*) En face view of granulated wound matrix 7 days after amputation. *Arrowheads* denote border of granule-encrusted structure within wound surface matrix. *Asterisks* denote collections of 50-nm granules, SEM. (*7b*) En face view of similar 'granulated' wound surface after digestion with chondroitinase AC II. *Arrows* denote fiber-like structures almost completely denuded of chondroitinase AC II labile 50-nm granules, SEM. *AEC*, apical epidermal cap; *B*, basal layers of regenerate epidermis; *C*, cartilage; *CB*, core blastema; *CT*, loose connective tissues; *D*, dermis; *E*, epithelial-like cells; *F*, free cell population; *FM*, fine filamentous material; *G/GM*, granulated wound surface matrix; *I*, inflammatory cells (macrophages and neutrophils); *N*, nerve; *RE*, regenerate epidermis; *SREB*, subregenerate epidermal blastema; *ST*, stump tissues; *TP*, trailing cells; *WE*, wound epidermis. (From Young et al. 1989aAnatomical Record 223:231–241, Reprinted with permission from Alan R. Liss, Inc.)

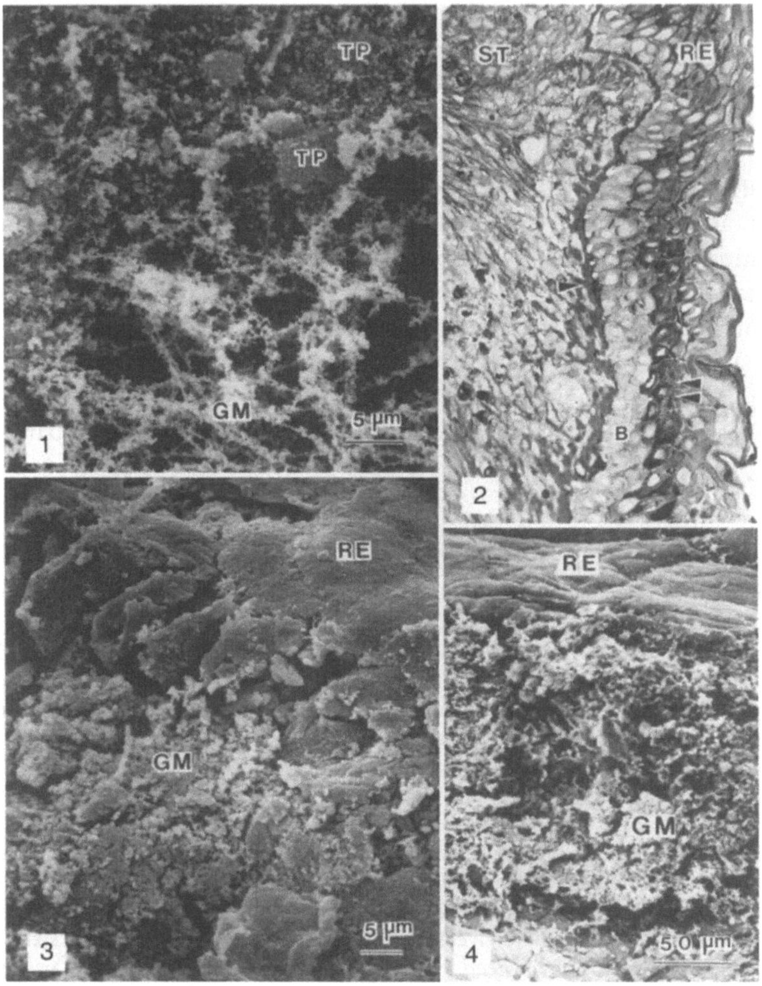

Fig. 4 (*1*) En face view of granulated wound surface matrix 7 days after amputation. Wound surface chondroitinase AC II-sensitive 50-nm granules co-localize with two populations of migrating cells. Free cell population 'pioneering' cells (first wave) co-localize with 50-nm granules located along fibers within matrix, lower half of field. Free cell population 'trailing' cells (second wave) co-localize with 50-nm granules located within the remaining spaces between the granule-encrusted fibers, upper half of field, resulting in almost complete granulation of matrix, SEM. (*2*) En face view of wound surface 10 days after amputation. Regenerate epidermal cuff begins 'purse-string'-like wound closure across outer surface of complete granulated (both along and between fibers within) wound surface matrix, SEM. (*3*) Histological view of wound closure by regenerate epidermal cuff across surface of stump tissues 10 days

sulfate proteoglycan and glycoproteins would then physically separate themselves from the connective tissues and migrate along a concentration gradient (Fig. 5, part 2) of glycoproteins and hyaluronic acid secreted by the regenerate epidermis towards the forming apical epidermal cap (Fig. 5, parts 4 and 5). This unique population of cells expressing cell surface-bound heparan sulfate proteoglycan and glycoproteins formed the mass of cells denoted as the sub-epidermal ridge blastema (SERB) and blastema (BL) (Fig. 5, parts 5 and 6). The blastema proliferated to increase the length of the limb. During this process, the more proximal blastema cells (near the differentiated tissues) acquired cell-surface coats consisting of combinations of acidic, neutral, and basic glycoproteins. The proteoglycans of the extracellular matrix then developed into tissue-specific profiles for skeletal muscle, fat, cartilage, bone, connective tissues, and dermis (Young et al. 1985). In the process of forming the missing tissues of the limb, the alterations in proteoglycan profiles preceded overt changes in cellular morphology. These results suggested that the blastema in the adult terrestrial salamander was formed by quiescent stem cells located within the more proximal connective tissues of the limb and that during differentiation the blastemal stem cells formed an intermediate tissue-specific (lineage-committed) phenotype before overt differentiation into the adult form.

after amputation. Intracellular staining for hyaluronic acid and glycoproteins occurs within middle layers of regenerate epidermis (*double arrowheads*), absent in basal layers of regenerate epidermis (*B*), and present extracellularly within adjacent underlying wound tissues (*single arrowhead*), ×1,200. (*4*) En face view of granulated wound surface matrix of original amputation site 15 days after amputation. The free cell population has reached the center of the wound surface. There is complete granulation of the wound surface matrix. There is also continued closure by the regenerate epidermal cuff across the outer surface of the granulated matrix to close the wound, SEM. *AEC*, apical epidermal cap; *B*, basal layers of regenerate epidermis; *C*, cartilage; *CB*, core blastema; *CT*, loose connective tissues; *D*, dermis; *E*, epithelial-like cells; *F*, free cell population; *FM*, fine filamentous material; *G/GM*, granulated wound surface matrix; *I*, inflammatory cells (macrophages and neutrophils); *N*, nerve; *RE*, regenerate epidermis; *SREB*, subregenerate epidermal blastema; *ST*, stump tissues; *TP*, trailing cells; *WE*, wound epidermis. (From Young et al. 1989a Anatomical Record 223:231–241, Reprinted with permission from Alan R. Liss, Inc.)

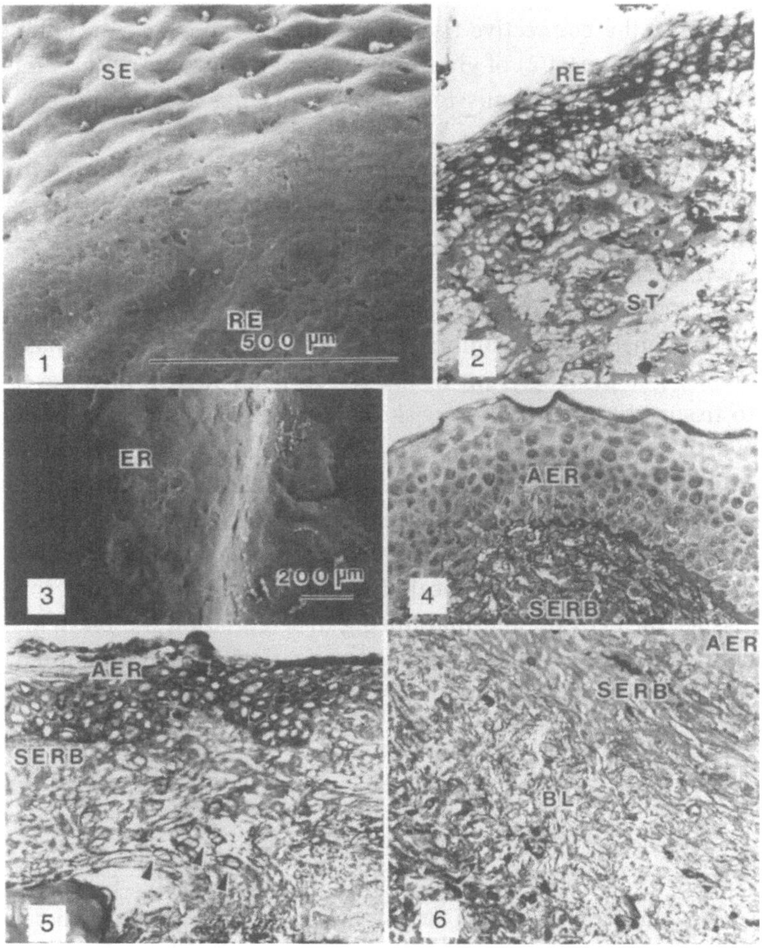

Fig. 5 (*1*) En face view of complete wound closure 20 days after amputation. The regenerate epidermis has completely covered the wound surface (lower field). The center of the wound is located at the lower right corner of the photograph. The stump epidermis continues to cover the lateral aspect of the limb (upper field). Note opening of pores for the mucous skin glands located within the stump epidermis and absent in the regenerate epidermis, SEM. (*2*) Regenerate epidermis overlying wound surface 22 days after amputation. An increase in intracellular staining for both hyaluronic acid and glycoproteins occurs in all layers of the regenerate epidermis. Increased extracellular staining for both hyaluronic acid and glycoproteins is located within underlying stump tissue matrices, ×550. (*3*) En face view of forming apical epidermal cap within regenerate epidermis covering wound site 25 days after amputation, SEM. (*4*) Distal regenerate portion of the stump–regenerate complex, sec-

3 Existence of Quiescent Stem Cells in Higher Order Animals

Previous studies demonstrated the existence of lineage-committed unipotent, bipotent, tripotent, and multipotent progenitor stem cells resident within differentiated tissues. For example, unipotent myosatellite myoblasts were reported for skeletal muscle (Mauro 1961; Campion 1984; Grounds et al. 1992), unipotent adipoblasts for adipose tissue (Ailhaud et al. 1992), unipotent chondrogenic cells and osteogenic cells for the perichondrium and periosteum, respectively (Cruess 1982), bipotent adipofibroblasts for adipose tissue (Vierck et al. 1996), bipotent chondrogenic/osteogenic cells for marrow (Owen 1988; Beresford 1989; Caplan et al. 1997; Prockop 1997), tripotent chondrogenic/osteogenic/adipogenic cells for marrow (Pittenger et al. 1999), and also multipotent hematopoietic cells for marrow (Palis and Segel 1998; McGuire 1998; Ratajczak et al. 1998).

tioned perpendicular to the preaxial/postaxial axis 25 days after amputation. Regenerate epidermis begins to thicken forming a 'cap' of epidermis covering the wound site. This structure is designated as the apical epidermal cap. The 'pioneering' cells and the 'trailing' cells become the cell population beneath the forming apical epidermal cap. This region is designated as the 'subregenerate epidermal blastema'. Hyaluronic acid is present intracellularly within all layers of the apical epidermal cap and extracellularly within the subregenerate epidermal blastema, ×600. (*5*) Stump–regenerate complex 30 days after amputation. All layers of the apical epidermal cap stain intracellular for hyaluronic acid and glycoproteins, while the subregenerate blastema stains extracellularly for the same material. Future blastemal cells (*arrowheads*) are visible near the subregenerate epidermal blastema. These blastemal cells stain intracellularly for heparan sulfate glycosaminoglycans, pericellularly for heparan sulfate glycosaminoglycans and glycoproteins, and extracellularly for hyaluronic acid, ×550. (*6*) Regenerate portion of stump regenerate complex 35 days after amputation. The individual mesenchymal-like cells have congregated into a large mass of cells, designated the core blastema, which lies deep (internal) to the subregenerate epidermal blastema, and the apical epidermal cap, ×600. *AEC*, apical epidermal cap; *B*, basal layers of regenerate epidermis; *C*, cartilage; *CB*, core blastema; *CT*, loose connective tissues; *D*, dermis; *E*, epithelial-like cells; *F*, free cell population; *FM*, fine filamentous material; *G/GM*, granulated wound surface matrix; *I*, inflammatory cells (macrophages and neutrophils); *N*, nerve; *RE*, regenerate epidermis; *SREB*, subregenerate epidermal blastema; *ST*, stump tissues; *TP*, trailing cells; *WE*, wound epidermis. (From Young et al. 1989a Anatomical Record 223:231–241, Reprinted with permission from Alan R. Liss, Inc.)

Based on previous studies with adult amphibians, Young and colleagues postulated the existence of stem cells that resided within the differentiated cells of tissues. These quiescent lineage-uncommitted stem cells had the capacity to form multiple cell types from different tissue lineages. They theorized that the technology for isolation and cultivation of these particular stem cells was unknown at that time, hence there were no previous reports for the existence of these putative pluripotent stem cells. After 3 years of empirical testing, optimal conditions were determined for the isolation, cultivation, and cryopreservation of putative pluripotent stem cells (Young et al. 1991, 1992a). A quantitative bioassay was also developed to objectively measure and quantify the biological activity of these cells resident stem cells (Young et al. 1992b).

To prove or disprove the pluripotency hypothesis necessitated the derivation of a population of cells derived from a single cell. This was accomplished by serial dilution clonogenic analysis, deriving multiple clonal populations from single cells (Young et al. 1993). A total of 1,158 clones of cells were generated in the initial cloning experiments. The original cells used for cloning were isolated from connective tissue compartments of skeletal muscle, dermis, and heart. All 1,158 clones generated displayed contact inhibition at confluence. Approximately 70% of the clones demonstrated multiple differentiation capabilities with respect to the mesodermal lineage. These clones formed skeletal muscle, smooth muscle, unilocular (white) fat, multilocular (brown) fat, hyaline cartilage, articular cartilage, elastic cartilage, growth plate cartilage, fibrocartilage, endochondral bone, and intramembranous bone (Young et al. 1993). The remaining 30% of the clones demonstrated tripotency (chondrogenic/osteogenic/adipogenic), bipotency (adipogenic/fibrogenic, myogenic/ fibrogenic, chondrogenic/osteogenic), or unipotency (adipogenic, myogenic, fibrogenic, chondrogenic, osteogenic) with respect to their inherent differentiation potentials. However, other differences between these two separate categories of stem cells were also noted.

The clones displaying either tripotency, bipotency or unipotency would proliferate through 30–50 cell doublings and then senesce and die. They were unresponsive to a general inductive agent (i.e., dexamethasone) or a specific inductive agent (i.e., bone morphogenetic protein-2; BMP-2) outside their tissue lineage. In other words, clones that would only form muscle tissues were unresponsive to BMP-2, which induces endochondral bone formation. However, these same clones were responsive to progression agents [such as insulin, insulin-like growth factor

(IGF)-1, and IGF-2] that would accelerate the time frame for expression of their particular cell type. For example, clones that would only form muscle tissues would demonstrate the muscle phenotype in 1 week in the presence of a progression agent, whereas 2 or more weeks would be required to express the muscle phenotype in the absence of a progression agent (Rogers et al. 1995; Young et al. 1998a).

In contrast, clones displaying pluripotency (i.e., the ability to form skeletal muscle, smooth muscle, unilocular fat, multilocular fat, hyaline cartilage, articular cartilage, elastic cartilage, growth plate cartilage, fibrocartilage, endochondral bone, and intramembranous bone from a single cloned cell) proliferated well past 200 cell doublings without any apparent effect on their differentiation potential. They were responsive to both a general inductive agent, such as dexamethasone, that would induce their full range of cell types, as well as being responsive to specific inductive agents, such as BMP-2, which would exclusively induce endochondral bone formation. However, these same clones were unresponsive to progression agents such as insulin, IGF-1, and IGF-2, which had no effect on the clones as long as they remained lineage-uncommitted. Once these multifunctional clones committed to a particular tissue lineage, they assumed all the characteristics of the tripotent, bipotent, and unipotent progenitor stem cell clones for that respective lineage. Their lifespan before senescence and death was 50–70 cell doublings. They were unresponsive to induction agents outside their lineage, but were responsive to progression agents. The clones with the capability to form multiple (more than four) distinct mesodermal cell types from a single cell were designated as '*pluripotent* mesenchymal stem cells' because of their ability to form multiple different tissues originating from the mesodermal germ layer. The clonal studies were repeated with putative pluripotent stem cells derived from mice and rats and similar results were noted (Rogers et al. 1995; Young et al. 1998a, 2001a).

4 Phylogenetic Distribution

We next hypothesized that pluripotent mesenchymal stem cells were present in higher order mammals. Thus far, putative quiescent stem cells with characteristics identical to the original pluripotent mesenchymal stem cell clones have been isolated from mice, rats, rabbits, dogs, sheep, goats, pigs, and humans (Table 1) (Pate et al. 1993; Lucas et al. 1995,;

Table 1 Distribution and age of donor tissue

Age	Avian	Mouse	Rat	Rabbit	Dog	Goat	Sheep	Pig	Human male	Human female
Fetal	+	+	+						24 weeks	25 weeks
Newborn		+	+						+	
Adolescent		+	+							+
Adult		+	+	+	+	+	+	+	37 years	17 years
									48 years	25 years
										36 years
										40 years
Geriatric		30months							67 years	77 years

Table 2 Source tissue for pluripotent stem cells

Source tissue	Avian	Mouse	Rat	Rabbit	Dog	Goat	Sheep	Pig	Human
Skeletal muscle CT	+	+	+	+	+	+	+	+	+
Dermis	+	+	+						+
Heart CT	+	+	+						
Granulation tissue			+						+
Other sites[a]	+		+						+

[a] Associated connective tissues of: periosteum, perichondrium, fat, ligaments, tendons, nerve sheaths, meninges, blood vessels, blood, bone marrow, trachea, esophagus, stomach, duodenum, jejunum, ileum, large intestine, liver, pancreas, spleen, kidney, and urinary bladder.
CT, connective tissue.

Rogers et al. 1995; Warejcka et al. 1996; Young et al. 1998a, 1999, 2001a,b, 2003a; Young and Black 2003).

5 Age of Donor

We next hypothesized that if the pluripotent mesenchymal stem cells were involved in repair and regeneration of tissues, the stem cells should be present at all stages of life. This hypothesis was examined in mice using the National Institute of Health's designated mouse aging model through 30 months of age and in humans through 77 years of age (Table 1). In both mice and humans no differences were detected with respect to the differentiation potentials of these cells. The differentiation potentials examined were the ability to induce phenotypic expression markers for skeletal muscle, smooth muscle, cardiac muscle, unilocular (white) fat, multilocular (brown) fat, hyaline cartilage, articular cartilage, elastic cartilage, growth plate cartilage, fibrocartilage, endochondral bone, intramembranous bone, tendon, ligament, dermis, scar tissue, endothelial cells, and hematopoietic in these cells using dexamethasone as a general inductive agent (Table 1) (Young et al. 1999, 2001b, 2003a; Young 2000; Young and Black 2003).

6 Location of Pluripotent Stem Cells

We next hypothesized that if the pluripotent mesenchymal stem cells were indeed involved in any repair and regeneration of tissues, the stem cells should be present throughout the body. Utilizing the same protocols, cells from 26 different tissues and organs were isolated and examined. Cells with the same characteristics as pluripotent mesenchymal stem cells were located in any tissue or organ with a connective tissue component (Table 2) (Young et al. 1991, 1995, 1999, 2001a,b, 2003a; H. E. Young, unpublished results; Pate et al. 1993; Lucas et al. 1995; Rogers et al. 1995; Warejcka et al. 1996; Young and Black 2003).

7
Effects of Bioactive Factors

We next hypothesized that if these 'pluripotent mesenchymal stem cells' were indeed involved in repair and regeneration of tissues, they should be refractive to agents and conditions known to be involved in generalized tissue replacement and repair. In a series of studies a large number of compounds were examined, both singly and in combination. We compared a mixed population of clonal lineage-committed progenitor stem cells (20% each of unipotent myoblasts, adipoblasts, chondroblasts, osteoblasts, and fibroblasts) with a single population of clonal pluripotent mesenchymal stem cells. Those studies demonstrated four basic types of biological activities expressed by the agents examined on the mixed progenitor stem cell populations and on the clonal pluripotent stem cells. Those activities were proliferative, progressive (accelerate phenotypic expression), inductive (commit cells to a particular tissue lineage), and/or inhibitory (prevention of progression and/or induction) to the generalized repair/regeneration response (Dixon et al. 1996; Lucas et al. 1996a,b; Young et al. 1998b; Young 2000).

8
Unique Cell Surface Antigens

We have also been involved in studies to accelerate the isolation of purified populations of pluripotent stem cells from the connective tissues of human subjects for potential use in bioengineered transplant therapies. We have been analyzing the cell surface cluster of differentiation (CD) markers (Kishimoto et al. 1997) coupled with flow cytometric analysis to determine if they could identify unique marker combinations on these stem cells. Thus far, 58 CD markers have been examined. Those markers are CD1a, CD2, CD3, CD4, CD5, CD7, CD8, CD9, CD10, CD11b, CD11c, CD13, CD14, CD15, CD16, CD18, CD19, CD20, CD22, CD23, CD24, CD25, CD31, CD33, CD34, CD36, CD38, CD41, CD42b, CD44, CD45, CD49d, CD55, CD56, CD57, CD59, CD61, CD62E, CD65, CD66e, CD68, CD69, CD71, CD79, CD83, CD90, CD95, CD105, CD117, CD123, FLT3 (CD135), CD166, Glycophorin-A, HLA Class-I, HLA-DRII, FMC-7, Annexin, and LIN antigens (Young et al. 1999, 2001b, 2003b). Six CD markers were identified that formed a unique 'fingerprint' for pluripotent stem cells using cells lines from human fetal, adult, and geriatric donors.

Those CD markers are CD10, CD13, CD34, CD56, CD90, and MHC-I (Young et al. 1999, 2001b).

9 Pluripotent Stem Cells Residing in Tissues

To analyze stem cells procedures designed specifically for the harvest of pluripotent stem cells from connective tissue compartments were used together with CD cell surface marker analysis. We had hypothesized that a single population of pluripotent mesenchymal stem cells would be present in adult differentiated tissues. Much to our surprise, we found five separate populations of putative pluripotent stem cells, each with their own unique 'fingerprint' of CD markers. We have been characterizing two of those five populations to date—the pluripotent mesenchymal stem cells (Young 2000; Young et al. 2001a,b) and pluripotent epiblastic-like stem cells (Young et al. 2003b,c).

10 Pluripotent Mesenchymal Stem Cells

The first population characterized expressed the CD10, CD13, CD34, CD56, CD90, and MHC-I cell-surface fingerprint. This population displayed all the characteristics seen previously for pluripotent mesenchymal stem cells, from both human cell lines and nonhuman clones. The stem cells are 10–20 μm in size (Fig. 6, part 1.1). The stem cells displayed extended capabilities for self-renewal well past Hayflick's limit, propagating past 200 cell doublings without loss of differentiative potential (Young 2000; Young et al. 2001a,b). The identified pluripotent mesenchymal stem cells were quiescent in a serum-free defined media without leukemia inhibitory factor or in the absence of a feeder layer. In the quiescent state they displayed no proliferation, no spontaneous differentiation, and no cell death. These stem cells displayed contact inhibition at confluence (Fig. 6, part 1.2) (Young et al. 2001a), ceasing cell proliferation once a single layer of cells covered the flask surface, even in the presence of known proliferation agents such as platelet-derived growth factors (Young et al. 1998a; Young 2000). The pluripotent mesenchymal stem cells were responsive to both general and specific inductive factors (Young et al. 1998a; Young 2000), demonstrating the ability to form 18 of more mesodermal phenotypes from a single cell (Young et al.

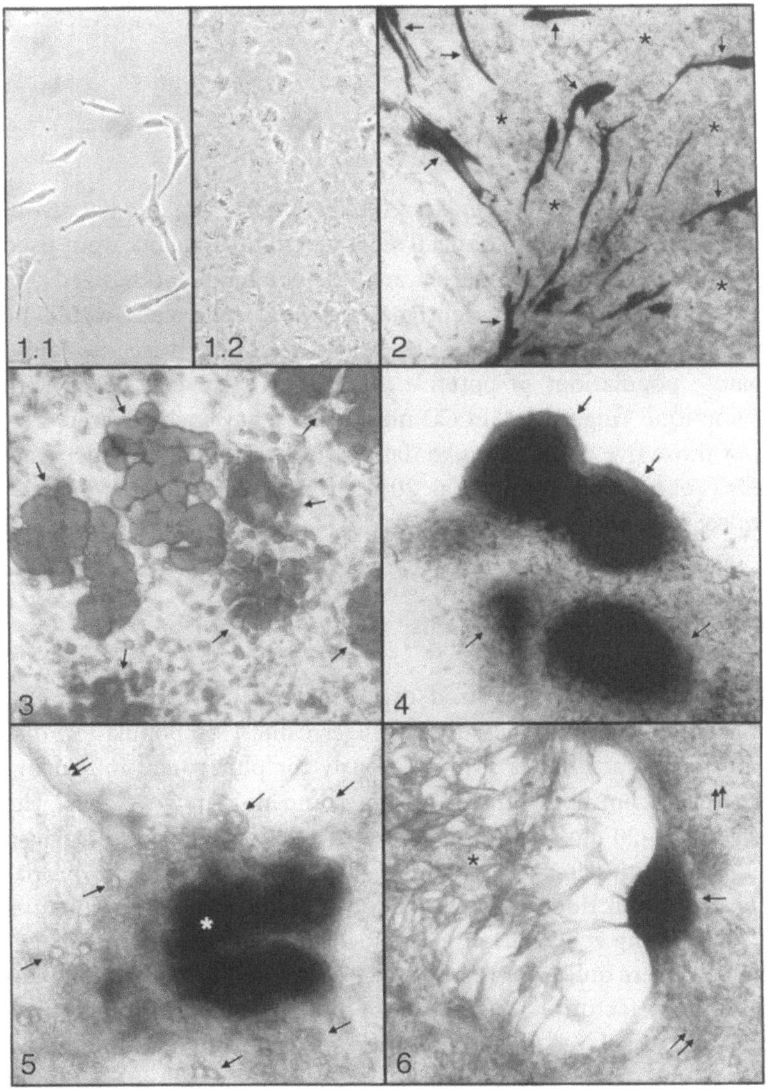

Fig. 6 Clone A2A2 propagated as noted and incubated for up to 47 days. Morphologies and histochemical and immunochemical staining as noted. (*1.1*) Cells 24 h after initial plating in SCM-1. Note presence of small mononucleated cells with large ratios of nucleus to cytoplasm. Unstained, phase-contrast, ×200. (*1.2*) Cells treated for 6 weeks in TM-6 with 2 μg/ml insulin. Note single layer of nondescript contact-inhibited cells. Unstained, phase contrast, ×200. (*2*) Cells treated for 6 weeks in TM-6 with 2 μg/ml insulin, HS, and 10^{-6} M dexamethasone. Note presence of dark struc-

2001a,b) (Table 3, Fig. 6, parts 2–6). These stem cells were unresponsive to progression factors (Young et al. 1998a; Young 2000) in the undifferentiated lineage-uncommitted state. The pluripotent mesenchymal stem cells would assume progenitor stem cell status once induced to commit to specific mesodermal tissue lineages (Table 3). Once committed to a particular tissue lineage, the newly induced lineage-specific progenitor stem cells would be unresponsive to inductive factors outside their respective lineage. Thus, stem cells committed to the myogenic lineage were unresponsive to BMP-2, and cells committed to the chondrogenic lineage were unresponsive to muscle morphogenetic protein (Young 2000). However, they responded to progression factors by accelerating the expression of their differentiated phenotype. The lineage-induced progenitor stem cells would also assume Hayflick's limit of 50–70 cell doublings before programmed cell senescence and death occurred (Young et al. 1999, 2001a,b; Young 2000).

11
Pluripotent Epiblastic-Like Stem Cells

The second population of putative pluripotent stem cells derived from adult human connective tissues expressed a CD10 and CD66e cell sur-

tures stained with antibody to sarcomeric myosin (MF-20). Majority of unstained cells in background are adipocytes (fat cells), ×100. (*3*) Cells treated for 6 weeks in TM-6 with 2 μg/ml insulin, HS, and 10^{-6} M dexamethasone. Note presence of cells with multiple Oil Red-O stained intracellular vesicles indicative of adipocytes, ×100. (*4*) Cells treated for 6 weeks in TM-6 with 2 μg/ml insulin, HS, and 10^{-6} M dexamethasone. Note presence of aggregating nodules of rounded cells with pericellular matrix halos staining with antibody to type-II collagen (CIIC1), ×100. (*5*) Cells treated for 6 weeks with 2 μg/ml insulin, HS, and 10^{-6} M dexamethasone. Note aggregating nodule of cells overlain with horse shoe-shaped extracellular matrix staining with antibody to bone sialoprotein (WV1D1). Unstained refractile vesicles belong to adipocytes. A diagonally oriented smooth muscle cell stained with an antibody to smooth muscle α-actin (1A4) is located in the upper left-hand corner of photograph, ×100. (*6*) Cells treated for 6 weeks with 2 μg/ml insulin, HS, and 10^{-6} M dexamethasone. Note aggregating nodule of cells (right side) with pericellular matrix halos staining with antibody to type-IX collagen (D1-9). Also note individual cells (left side) stained with antibody to smooth muscle α-actin (1A4), ×100. (From Young et al. 2001a, Anatomical Record 263:350–360, reprinted with permission from Wiley-Liss, Inc.)

Table 3 Induction of phenotypic expression in postnatal pluripotent stem cells

Phenotypic markers	PPELSCs	PPEctoSCs	PPMSCs	PPEndoSCs
Embryonic				
Alkaline phosphatase	+	–	–	–
SSEA-1	+	–	–	–
SSEA-3	+	–	–	–
SSEA-4	+	–	–	–
CEA	+	–	–	–
HCEA	+	–	–	–
CD66e	+	–	–	–
CEA-CAM	+	–	–	–
Ectoderm				
Neuronal progenitor cells	+	+	–	–
Neurons	+	+	–	–
Ganglia	+	+	–	–
Oligodendrocytes	+	+	–	–
Astrocytes	+	+	–	–
Radial glial cells	+	+	–	–
Keratinocytes	+	+	–	–
Mesoderm				
Skeletal muscle	+	–	+	–
Smooth muscle	+	–	+	–
Cardiac muscle	+	–	+	–
Unilocular (white) fat	+	–	+	–
Multilocular (brown) fat	+	–	+	–
Hyaline Cartilage	+	–	+	–
Articular cartilage	+	–	+	–
Elastic cartilage	+	–	+	–
Growth plate cartilage	+	–	+	–
Fibrocartilage	+	–	+	–
Endochrondral bone	+	–	+	–
Intramembranous bone	+	–	+	–
Tendon	+	–	+	–
Ligament	+	–	+	–
Dermis	+	–	+	–
Scar tissue	+	–	+	–
Endothelial cells	+	–	+	–
Erythroid cells	+	–	+	–
Myeloid cells	+	–	+	–
Endoderm				
Endodermal progenitor cells	+	–	–	+
Gastrointestinal epithelium	+	–	–	+
Liver oval cells	+	–	–	+

Table 3 (continued)

Phenotypic markers	PPELSCs	PPEctoSCs	PPMSCs	PPEndoSCs
Liver biliary cells	+	–	–	+
Liver canalicular cells	+	–	–	+
Liver hepatocytes	+	–	–	+
Pancreatic progenitor cells	+	–	–	+
Pancreatic ductal cells	+	–	–	+
Pancreatic α-cells	+	–	–	+
Pancreatic β-cells	+	–	–	+
Pancreatic δ-cells	+	–	–	+

PPELSCs, pluripotent epiblastic-like stem cells (isolated and cloned); PPEctoSCs, pluripotent ectodermal stem cells (induced); PPMSCs, pluripotent mesenchymal stem cells (isolated and cloned); PPEndoSCs, pluripotent endodermal stem cells (induced) (Young et al. 2001a, 2001b, 2002a, 2002b); SSEA-1, stage-specific embryonic antigen-1; CEA, carcinoembryonic antigen; HCEA, human carcinoembryonic antigen; CD66e, carcinoembryonic antigen; CEA-CAM, carcinoembryonic antigen-cell adhesion molecule (Young et al. 2002a, 2002b).

face fingerprint. We have also cloned a putative pluripotent stem cell population with similar characteristics from postnatal rats. These cells share similarities and differences with the pluripotent mesenchymal stem cells. Similarities with pluripotent mesenchymal stem cells include extended capabilities for self-renewal, quiescence in a serum-free defined media (without leukemia inhibitory factor or in the absence of a feeder layer), responsiveness to inductive factors, unresponsiveness to progression factors, and assumption of progenitor stem cell status once committed to a particular tissue lineage. The progenitor stem cell exhibits contact inhibition at confluence, lack of responsiveness to induction factors outside its respective tissue lineage, responsiveness to progression factors accelerating its phenotypic expression, and conformance to Hayflick's limit of 50–70 cell doublings before programmed cell senescence and death occurred. However, differences between these two cell populations strongly suggested the existence of unique populations of pluripotent stem cells.

The second pluripotent stem cell population derived from adults is 6–8 mm in size (Fig. 7, parts 4–7; Fig. 8, part 1). They express embryonic stem cell markers (i.e., alkaline phosphatase, stage-specific embryonic antigen-1, -3, -4, and carcinoembryonic antigens) in a quiescent state (Fig. 7, part 6 and 7; Fig. 8, parts 2–4). The stem cells did not exhibit contact inhibition at confluence, but rather formed multiple confluent

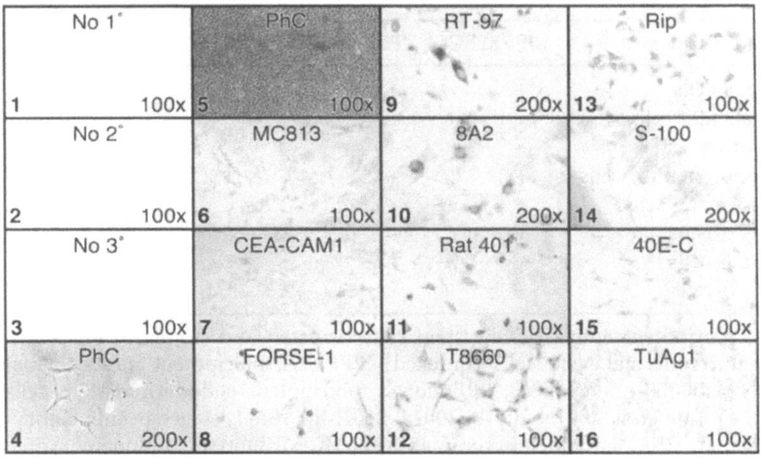

Fig. 7 Clone Rat-A2B2 grown for 24 h to 8 weeks as designated in serum-free testing medium; testing medium and insulin (2 μg/ml); testing medium and inductive agents (i.e., 10^{-10} to 10^{-6} M dexamethasone or selected sera), testing medium, insulin, and dexamethasone; or testing medium, insulin, dexamethasone, and selected sera. Morphologies and histochemical and immunochemical staining as noted. Photographed with brightfield microscopy unless otherwise noted. Original magnifications as noted on photographs. (*1*) Representative staining control, no primary antibody. Cells grown for 2 weeks in testing medium plus 10^{-8} M dexamethasone. Primary antibody was replaced with Dulbecco's phosphate buffered saline, pH 7.4. (*2*) Representative staining control, no secondary antibody. Cells grown for 2 weeks in testing medium plus 10^{-8} M dexamethasone. Primary antibody was MY-32. Secondary antibody was replaced with Dulbecco's phosphate buffered saline, pH 7.4. (*3*) Representative staining control, no tertiary probe. Cells grown for 2 weeks in testing medium plus 10^{-8} M dexamethasone. Primary antibody was MY-32. Tertiary probe was replaced with Dulbecco's phosphate buffered saline, pH 7.4. (*4*) Cells grown for 24 h in plating medium. Note small stellate cells with large nuclear/cytoplasm ratios. Phase contrast microscopy. (*5*) Cells grown for 3 weeks in testing medium plus insulin. Note multiple layers of confluent cells. Phase contrast microscopy. (*6*) Cells grown for 1 week in testing medium plus insulin. Note mononucleated cells staining for stage-specific embryonic antigen-4 (MC813). (*7*) Cells grown for 1 week in testing medium plus insulin. Note mononucleated cells staining for carcinoembryonic antigen-cell adhesion molecule-1 (CEA-CAM1). (*8*) Cells grown for 1 week in testing medium, insulin, 10^{-6} M dexamethasone, and selected sera. Note mononucleated cells staining for neural precursor cell expression marker (FORSE-1). (*9*) Cells grown for 1 week in testing medium, insulin, 10^{-6} M dexamethasone, and selected sera. Note mononucleated cells showing intracellular staining for neurofilaments (RT-97). (*10*) Cells grown for 1 week in testing medium, insulin, 10^{-6} M dexamethasone, and selected sera. Note mononucleated cells showing intracellular staining for neurons (8A2). (*11*) Cells grown for 1 week in testing medium, insulin, 10^{-6} M dexamethasone, and selected sera. Note mononucleated cells showing intracellular staining for

layers in vitro (Fig. 7, part 5). These stem cells formed over 36 distinct cell types from all three primary germ layers: seven or more ectodermal cell types (i.e., neuronal progenitor cells, neurons, ganglia, oligodendrocytes, astrocytes, radial glial cells, keratinocytes) (Table 3, Fig. 7, parts 8–16; Fig. 8, parts 5–9), 18 or more mesodermal cell types (i.e., skeletal muscle, cardiac muscle, smooth muscle, white fat, brown fat, hyaline cartilage, elastic cartilage, growth plate cartilage, articular cartilage, fibrocartilage, cortical bone, trabecular bone, loose fibrous connective tissues, tendon, ligament, scar-connective tissue, endothelial cells, hematopoietic cells) (Table 1, Fig. 8, parts 11–16; Fig. 9, parts 1–28), and 11 or more endodermal cell types (i.e., endodermal progenitor cells, gastrointestinal epithelium, liver oval cells, liver hepatocytes, liver biliary cells, liver canalicular cells, pancreatic progenitor cells, pancreatic ductal cells, and glucagon-secreting α-cells, insulin-secreting β-cells, and somatostatin-secreting δ-cells of pancreatic islets) (Table 3, Fig. 7, parts 17 and18; Fig. 10, parts 1–18). We proposed that this second population of stem cells were a sequestered embryonic-like stem cell existing within an adult. We have designated this population as a 'pluripotent epiblastic-like stem cell', due to its ability to form cell types from all three primary germ layers (Young et al. 2003b, 2003c).

12
Capability for Extended Self-Renewal

One of the hallmarks of embryonic stem cells is their capability for extended self-renewal (Thomson et al. 1998; Shamblott et al. 1998), en-

neuronal nestin (Rat-401). (*12*) Cells grown for 1 week in testing medium, insulin, 10^{-6} M dexamethasone, and selected sera. Note mononucleated cells showing intracellular staining for β-tubulin-III (T8660). (*13*) Cells grown for 1 week in testing medium, insulin, 10^{-6} M dexamethasone, and selected sera. Note mononucleated cells showing intracellular staining for rat-specific oligodendrocytes (Rip). (*14*) Cells grown for 1 week in testing medium, insulin, 10^{-6} M dexamethasone, and selected sera. Note mononucleated cells showing intracellular staining for neuronal expression marker (S-100). (*15*) Cells grown for 1 week in testing medium, insulin, 10^{-6} M dexamethasone, and selected sera. Note mononucleated cells showing intracellular staining for neuronal vimentin for radial cells and radial glial cells (40E-C). (*16*) Cells grown for 1 week in testing medium, insulin, 10^{-6} M dexamethasone, and selected sera. Note mononucleated cells showing intracellular staining for ganglion cells (TuAg1). (From Young et al. 2003b)

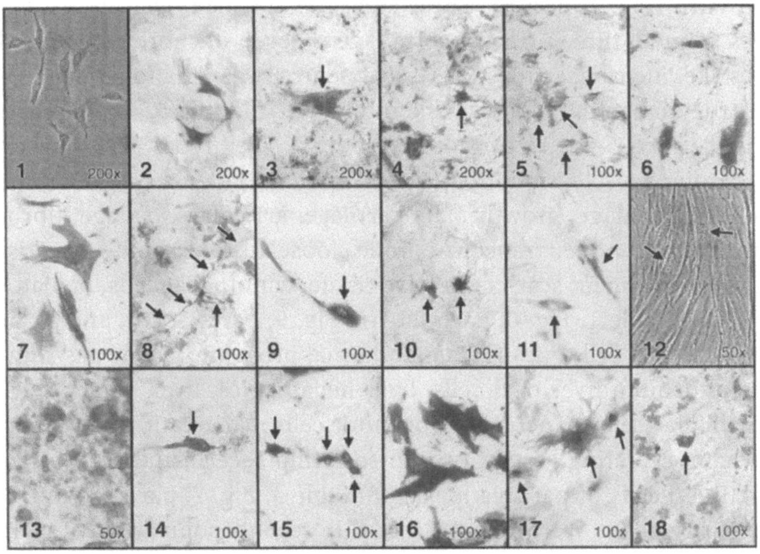

Fig. 8 NHDF-2 cells incubated with control medium plus dexamethasone for 56 days, unless otherwise noted. Cells were photographed using either phase contrast (Ph) or brightfield (BF) microscopy at original magnification noted. (*1*) Cells incubated for 24 h after plating. Note eight very small cells with high nuclear/cytoplasm ratios, Ph, ×200. (*2*) Note two very small cells heavily stained with antibody to stage-specific embryonic antigen-1 (MC480), BF, ×200. (*3*) Note one very small cell (*arrow*) stained with antibody to stage-specific embryonic antigen-3 (MC631), BF, ×200. (*4*) Note one very small cell (*arrow*) heavily stained with antibody to stage-specific embryonic antigen-4 (MC813-70), BF, ×200. (*5*) Note four cells (*arrows*) stained with antibody to nestin (MAB353), BF, ×200. (*6*) Note four cells stained with antibody to neurons (S-100), BF, ×100. (*7*) Note multiple cells stained with antibody to neurofilaments (RT-97), BF, ×100. (*8*) Note single cell with long cell processes (*arrows*) stained with antibody to neurofilaments (N-200), BF, ×100. (*9*) Note single cell stained with antibody for neuroglia (CNPase), BF, ×100. (*10*) Note two cells (*arrows*) heavily stained with antibody for keratinocytes (VM-1), BF, ×100. (*11*) Note two cells (*arrows*) stained with antibody to myogenin (F5D), BF, ×100. (*12*) Note two structures (*arrows*) containing multiple nuclei, Ph, ×50. (*13*) Note multiple cells containing Oil Red-O stained intracellular vesicles, BF, ×100. (*14*) Note single cell stained with antibody to type-II collagen (CIIC1), BF, ×100. (*15*) Note four cells (*arrows*) stained intracellularly with antibody to bone sialoprotein-II (WV1D1), BF, ×100. (*16*) Note multiple cells staining with an antibody to peripheral cell adhesion molecule (PE-CAM, P2B1), BF, ×100. (*17*) Note three cells (*arrows*) with intracellular vesicles stained for antibody to human-specific α-fetoprotein (HAFP), BF, ×100. (*18*) Note single cell (*arrow*) heavily stained with antibody to human-specific gastrointestinal epithelial-specific antigen (HESA), BF, ×100. (From Young et al. 2003b)

abling them to proliferate past Hayflick's limit of 50–70 cell doublings (Hayflick 1965). We hypothesized that if the pluripotent stem cells were similar to embryonic stem cells in their ability to propagate past Hayflick's limit, then they should be telomerase positive. To date, we have examined nonhuman clones of both the pluripotent mesenchymal stem cells and pluripotent epiblastic-like stem cells. Both sets of pluripotent clones display positive telomerase activity (Fig. 11) (Young et al. 2003c). The presence of telomerase activity is consistent with the capability for extended self-renewal exhibited by pluripotent mesenchymal stem cells and pluripotent epiblastic-like stem cells.

13
Paradigm Shift

Recently there has been an apparent paradigm shift regarding stem cell biology and embryogenesis—particularly with regards to stem cells derived from adult tissues. In the last few years, researchers are reporting stem cells derived from one type of adult differentiated tissue forming another type of adult differentiated tissue. For example, stem cells derived from the bone marrow have been reported to form neurons and neural supportive tissues (Eglitis and Mezey 1997; Woodbury et al. 2000). Stem cells derived from neuronal tissues have been reported to form blood elements (Bjornson et al. 1999; Vescovi et al. 2001), and stem cells derived from skeletal muscle have been reported to form blood (Jackson et al. 1999). These investigators have suggested that the stem cells from one organ can be reprogrammed, or transdifferentiate, to form cells and tissues of another organ.

Our studies support a different hypothesis to account for these results (Young 1977a–c, 1983, 2000; Young et al. 1983a–c, 1985, 1989a,b, 1991, 1992a,b, 1993, 1995, 1998a,b, 2001a,b, 2003a–c; Lucas et al. 1993, 1995, 1996a,b, P.A. Lucas, unpublished results; Pate et al. 1993; Rogers et al. 1995; Dixon et al. 1996; Warejcka et al. 1996; Young and Black, 2003). We observed that the connective tissue compartments of many organs and tissues contain a variety of different categories of stem cells. This concept has been substantiated through the use of histochemical analyses, clonogenic analyses, isolation studies, cell sorting studies, and phenotypic expression studies across nine different species, including humans. These studies demonstrated that approximately 50% of the resident stem cells in any given connective tissue compartment are lineage-committed

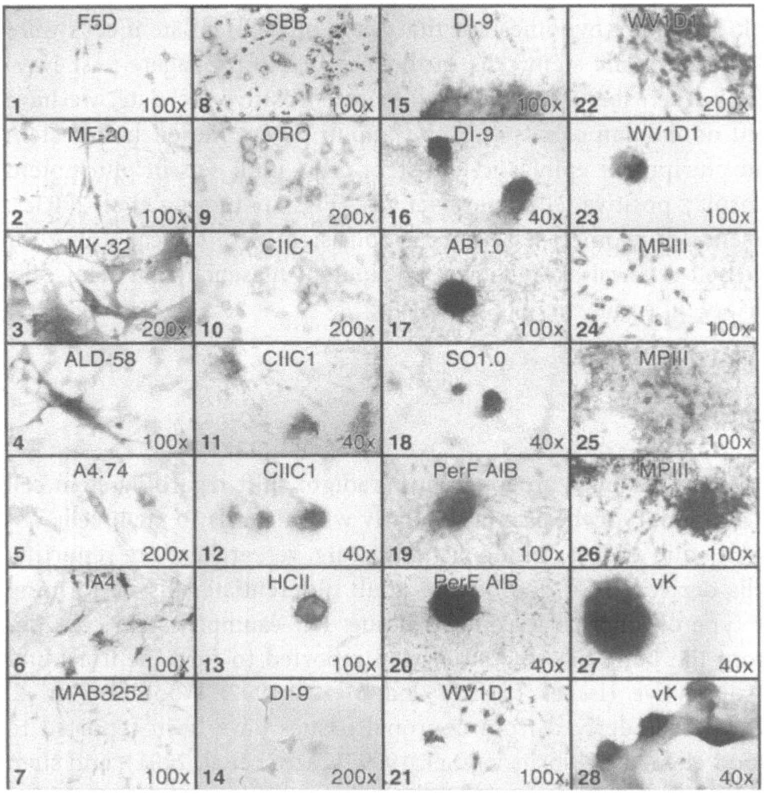

Fig. 9 Clone Rat-A2B2 grown for 24 h to 8 weeks as designated in testing medium and inductive agents (i.e., 10^{-10} to 10^{-6} M dexamethasone or selected sera), testing medium, insulin, and dexamethasone; or testing medium, insulin, dexamethasone, and selected sera. Morphologies and histochemical and immunochemical staining as noted. Photographed with brightfield microscopy. Original magnifications as noted on photographs. (*1*) Clone grown for 1 week in testing medium, insulin, and 10^{-8} M dexamethasone. Note mononucleated cells showing intracellular staining for myogenin (F5D). (*2*) Clone grown for 2 weeks in testing medium, insulin, and 10^{-8} M dexamethasone. Note mononucleated cells showing intracellular staining for sarcomeric myosin (MF-20). (*3*) Clone grown for 2 weeks in testing medium, insulin, and 10^{-8} M dexamethasone. Note mononucleated cells showing intracellular staining for anti-skeletal muscle fast myosin (MY-32). (*4*) Clone grown for 2 weeks in testing medium, insulin, and 10^{-8} M dexamethasone. Note mononucleated cells showing intracellular staining for skeletal myosin heavy chain (ALD58). (*5*) Clone grown for 2 weeks in testing medium, insulin, and 10^{-8} M dexamethasone. Note mononucleated cells showing intracellular staining for skeletal myosin fast chain (A4.74). (*6*) Clone grown for 2 weeks in testing medium, insulin, and 10^{-8} M dexamethasone. Note mononucleated cells showing intracellular staining for smooth muscle ?-actin (IA4). (*7*) Clone grown for 1 week in testing medium, insulin, and 10^{-8} M dexamethasone. Note mononucleated cells show-

ing intracellular staining for cardiotin (cardiac myocytes, MAB 3252). (*8*) Clone grown for 2 weeks in testing medium, insulin, and 10^{-8} M dexamethasone. Note mononucleated cells with intracellular vesicles stained for saturated neutral lipids (SBB). (*9*) Clone grown for 2 weeks in testing medium, insulin, and 10^{-8} M dexamethasone. Note mononucleated cells with intracellular vesicles stained for saturated neutral lipids (ORO). (*10*) Clone grown for 2 weeks in testing medium, insulin, and 10^{-7} M dexamethasone. Note mononucleated cells with intracellular staining for cartilage-specific collagen pro type-II (CIIC1). (*11*) Clone grown for 3 weeks in testing medium, insulin, and 10^{-7} M dexamethasone. Note aggregations of cells demonstrating extracellular staining for cartilage-specific collagen pro type-II (CIIC1). (*12*) Clone grown for 4 weeks in testing medium, insulin, and 10^{-7} M dexamethasone. Note nodules of cells demonstrating extracellular staining for cartilage-specific collagen pro type-II (CIIC1). (*13*) Clone grown for 4 weeks in testing medium, insulin, and 10^{-7} M dexamethasone. Note a single nodule of cells demonstrating extracellular staining for cartilage-specific collagen type-II (HC-II). (*14*) Clone grown for 2 weeks in testing medium, insulin, and 10^{-7} M dexamethasone. Note mononucleated cells demonstrating intracellular staining for cartilage-specific collagen type-IX (D1-9). (*15*) Clone grown for 3 weeks in testing medium, insulin, and 10^{-7} M dexamethasone. Note aggregations of cells demonstrating extracellular staining for cartilage-specific collagen type-IX (D1-9). (*16*) Clone grown for 4 weeks in testing medium, insulin, and 10^{-7} M dexamethasone. Note two nodules of cells demonstrating extracellular staining for cartilage-specific collagen type-IX (D1-9). (*17*) Clone grown for 6 weeks in testing medium, insulin, and 10^{-7} M dexamethasone. Note single nodule of cells demonstrating extracellular staining for sulfated glycosaminoglycan chains of proteoglycans (AB1.0). (*18*) Clone grown for 6 weeks in testing medium, insulin, and 10^{-7} M dexamethasone. Note two nodules of cells demonstrating extracellular staining for sulfated glycosaminoglycan chains of proteoglycans (SO1.0). (*19*) Clone grown for 4 weeks in testing medium, insulin, and 10^{-7} M dexamethasone. Note single nodule of cells demonstrating extracellular staining for sulfated glycosaminoglycan chains of proteoglycans (PerFAlB). (*20*) Clone grown for 6 weeks in testing medium, insulin, and 10^{-7} M dexamethasone. Note single nodule of cells demonstrating extracellular staining for sulfated glycosaminoglycan chains of proteoglycans (PerFAlB). (*21*) Clone grown for 1 week in testing medium, insulin, and 10^{-8} M dexamethasone. Note mononucleated cells demonstrating intracellular staining for bone sialoprotein II (WV1D1). (*22*) Clone grown for 3 weeks in testing medium, insulin, and 10^{-8} M dexamethasone. Note aggregations of cells demonstrating extracellular staining for bone sialoprotein II (WV1D1). (*23*) Clone grown for 6 weeks in testing medium, insulin, and 10^{-8} M dexamethasone. Note nodule of cells demonstrating extracellular staining for bone sialoprotein II (WV1D1). (*24*) Clone grown for 1 week in testing medium, insulin, and 10^{-8} M dexamethasone. Note mononucleated cells demonstrating intracellular staining for osteopontine (MP111). (*25*) Clone grown for 3 weeks in testing medium, insulin, and 10^{-8} M dexamethasone. Note aggregations of cells demonstrating extracellular staining for osteopontine (MP111). (*26*) Clone grown for 6 weeks in testing medium, insulin, and 10^{-8} M dexamethasone. Note forming nodule of cells demonstrating extracellular staining for osteopontine (MP111). (*27*) Clone grown for 8 weeks in testing medium, insulin, and 10^{-8} M dexamethasone. Note nodule of cells demonstrating extracellular staining for calcium phosphate using the von Kossa procedure (vK). (*28*) Clone grown for 8 weeks in testing medium, insulin, and 10^{-8} M dexamethasone. Note four nodules of cells demonstrating extracellular staining for calcium phosphate using the von Kossa procedure (vK). (From Young et al. 2003b)

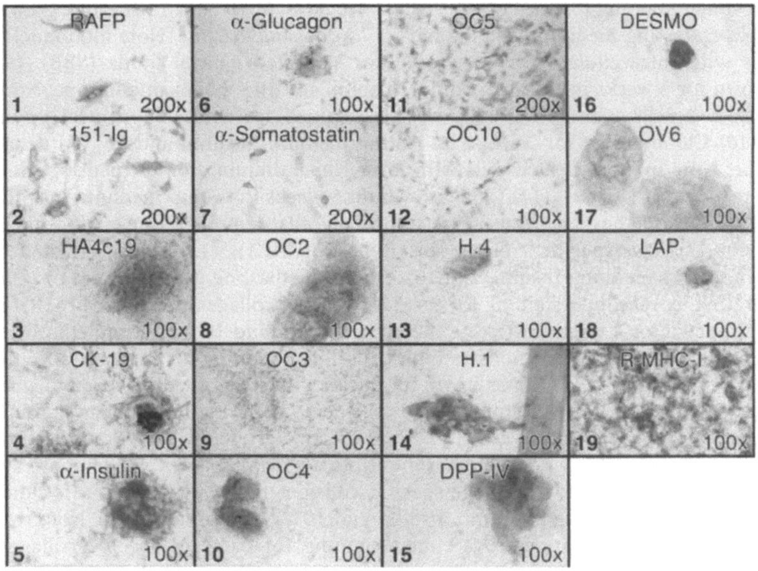

Fig. 10 Clone Rat-A2B2 grown for 24 h to 8 weeks as designated in testing medium and inductive agents (i.e., 10^{-10} to 10^{-6} M dexamethasone or selected sera), testing medium, insulin, and dexamethasone; or testing medium, insulin, dexamethasone, and selected sera. Morphologies and histochemical and immunochemical staining as noted. Photographed with brightfield microscopy. Original magnifications as noted on photographs. (*1*) Clone grown for 1 week in testing medium, insulin, 10^{-6} M dexamethasone, and selected sera. Note mononucleated and binucleated cells showing intracellular staining for rat-specific α-fetoprotein (RAFP). (*2*) Clone grown for 1 week in testing medium, insulin, 10^{-6} M dexamethasone, and selected sera. Note mononucleated and binucleated cells showing intracellular staining for rat-specific liver epithelial growth factor receptor (151-Ig). (*3*) Clone grown for 4 weeks in testing medium, insulin, 10^{-6} M dexamethasone, and selected sera. Note nodular aggregation cells showing intracellular staining for bile canalicular cells of liver (HA4c19). (*4*) Clone grown for 3 weeks in testing medium, insulin, 10^{-6} M dexamethasone, and selected sera. Note nodular aggregation cells showing intracellular staining for ductal cells of exocrine pancreas (CK-19). (*5*) Clone grown for 3 weeks in testing medium, insulin, 10^{-6} M dexamethasone, and selected sera. Note nodular aggregation cells showing intracellular staining for pro-insulin of endocrine pancreas (β-cells). (*6*) Clone grown for 2 weeks in testing medium, insulin, 10^{-6} M dexamethasone, and selected sera. Note nodular aggregation cells showing a few cells with intracellular staining for glucagon of endocrine pancreas (α-cells). (*7*) Clone grown for 1 week in testing medium, insulin, 10^{-6} M dexamethasone, and selected sera. Note individual mononucleated cells showing intracellular staining for somatostatin of endocrine pancreas (δ-cells). (*8*) Clone grown for 4 weeks in testing medium, insulin, 10^{-6} M dexamethasone, and selected sera. Note aggregation cells showing intracellular staining for progenitor cells, biliary epithelial cells, and oval cells of liver (OC2). (*9*) Clone

progenitor cells for that tissue. Approximately 40% of the resident stem cells are lineage-committed unipotent, bipotent, tripotent, and multipotent progenitor cells for other tissue lineages. Approximately 9% of the resident stem cells are pluripotent stem cells belonging to the ectodermal, mesodermal, or endodermal cell lineages, and approximately 1% of the resident stem cells are pluripotent epiblastic-like stem cells, capable of forming any somatic cell of the body. Using the stem cells resident within skeletal muscle connective tissue as an example, approximately 50% consist of progenitor stem cells committed to the myogenic lineage, approximately 40% are progenitor stem cells committed to other lineages (i.e., fibroblasts, adipoblasts, osteoblasts, chondroblasts, endothe-

◀

grown for 3 weeks in testing medium, insulin, 10^{-6} M dexamethasone, and selected sera. Note diffuse and aggregated cells showing intracellular staining for progenitor cells and biliary epithelial cells of liver (OC3). (*10*) Clone grown for 5 weeks in testing medium, insulin, 10^{-6} M dexamethasone, and selected sera. Note aggregated cells showing intracellular staining for progenitor cells and biliary epithelial cells of liver (OC4). (*11*) Clone grown for 2 weeks in testing medium, insulin, 10^{-6} M dexamethasone, and selected sera. Note diffuse cells showing intracellular staining for progenitor cells and biliary epithelial cells of liver (OC5). (*12*) Clone grown for 2 weeks in testing medium, insulin, 10^{-6} M dexamethasone, and selected sera. Note diffuse cells showing intracellular staining for progenitor cells and biliary epithelial cells of liver (OC10). (*13*) Clone grown for 3 weeks in testing medium, insulin, 10^{-6} M dexamethasone, and selected sera. Note diffuse and aggregated cells showing intracellular staining for cytoplasm of liver hepatocytes (H.4). (*14*) Clone grown for 3 weeks in testing medium, insulin, 10^{-6} M dexamethasone, and selected sera. Note diffuse and aggregated cells showing intracellular staining for liver hepatocyte cell surface marker (H.1). (*15*) Clone grown for 4 weeks in testing medium, insulin, 10^{-6} M dexamethasone, and selected sera. Note aggregated cells showing intracellular staining for progenitor cells, canalicular cells, and biliary epithelial cells of liver (DPP-IV). (*16*) Clone grown for 5 weeks in testing medium, insulin, 10^{-6} M dexamethasone, and selected sera. Note aggregated nodule of cells showing intracellular staining for endodermal epithelial marker of liver (DESMO). (*17*) Clone grown for 4 weeks in testing medium, insulin, 10^{-6} M dexamethasone, and selected sera. Note aggregated cells showing intracellular staining for biliary epithelial cells, oval cells, and hepatocellular carcinomaof liver (OV6). (*18*) Clone grown for 3 weeks in testing medium, insulin, 10^{-6} M dexamethasone, and selected sera. Note aggregated cells showing intracellular staining for canalicular cell surface protein of liver (LAP). (*19*) Clone grown for 2 weeks in testing medium, insulin, 10^{-6} M dexamethasone, and selected sera. Note diffuse cells showing intracellular staining for rat-specific MHC Class-I marker of differentiated rat cells (R-MHC-I). (From Young et al. 2003b)

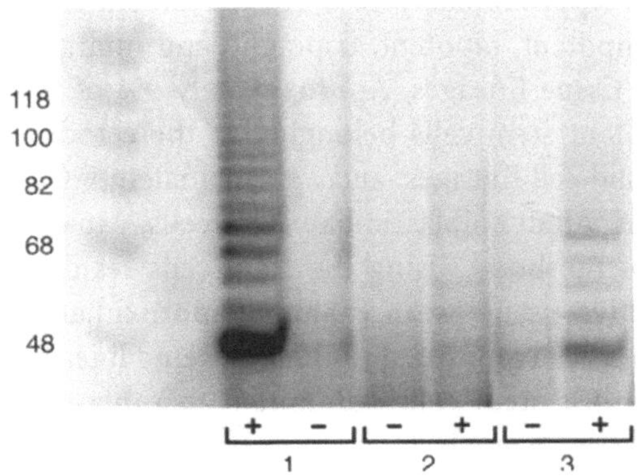

Fig. 11 Polyacrylamide gel electrophoresis of telomerase assays of clonal populations of rat pluripotent stem cells. *Lane 1+*, extract of telomerase positive cells (control), *lane 1–*, extraction buffer; *lanes 2 and 3 +*, test extract, *lanes 2 and 3–*, heat inactivated extract; *lane 2*, Rat-A2A2-scl-2PG (PPMSCs) at 31 passages and 151 cell doublings; and *lane 3*, Rat-A2B2-scl-40 (pluripotent epiblastic-like stem cells, PPELSCs) at 52 passages and 234 cell doublings. Note the presence of ladders of bands denoting the presence of telomerase activity, compare + *lanes 1–3*. (From Young et al. 2003b)

lioblasts, hematoblasts, neuroblasts, glioblasts, etc.), approximately 9% are pluripotent ectodermal stem cells, pluripotent mesodermal stem cells, and pluripotent endodermal stem cells, and approximately 1% are pluripotent epiblastic-like stem cells. Stem cells from the connective tissue compartments of such diverse tissues as dermis, fat, perichondrium, periosteum, bone marrow, brain, and viscera appear to follow the same pattern of distribution (Young et al. 1993, 1995, 2003a; Young 2000; Young and Black 2003).

Our work suggests that the reports supporting cell reprogramming (i.e., transdifferentiation) could be also explained by the presence of pluripotent stem cells (and/or lineage-committed progenitor stem cells for other tissue lineages) within the differentiated tissues. For example, we would propose that pluripotent stem cells (and/or hematopoietic progenitor cells) in skeletal muscle could give rise to blood cells (Jackson et al. 1999). Indeed, McKinney-Freeman et al. (2002), in an exquisite study, demonstrated that $CD45^+$ (hematopoietic) stem cells residing in skeletal muscle gave rise to blood cells, whereas the CD45– (nonhematopoietic) stem cells residing in muscle formed muscle. Similarly pluripotent stem cells (and/or neuroblasts and glioblasts) in bone marrow could give rise to neurons and neural supportive tissues (Eglitis and Mezey 1997; Woodbury et al. 2000). Likewise, pluripotent stem cells (and/or hematopoietic progenitor cells) in neuronal tissues could form blood elements (Bjornson et al. 1999; Vescovi et al. 2001). Therefore, no reprogramming of either differentiated cells or lineage-committed stem cells is required to explain the results obtained by these investigators.

14
Activation of Reserve Quiescent Stem Cells

Our studies support the hypothesis that the activation of reserve quiescent stem cells is a prerequisite for tissue restoration. Thus, both lineage-committed progenitor stem cells and lineage-uncommitted pluripotent stem cells are present throughout the individual as reserve populations of stem cells. During tissue replacement these quiescent stem cells become activated, proliferate, and differentiate into the missing tissues.

We propose that as development of an individual proceeds from the totipotent zygote, cells proliferate and segregate by lineage-commitment into two pluripotent layers, the trophoblast, which eventually forms the extra-embryonic membranes of the placenta, and the inner cell mass, which will eventually form the embryo. The inner cell mass continues its development by segregating into the hypoblast and epiblast. These layers further segregate into the pluripotent primary germ layers of the ectoderm, mesoderm, and endoderm. Further segregation through lineage-commitment into multipotent, tripotent, bipotent and eventually unipotent lineages further defines the differentiation pathways of the cells and their ultimate function.

Although a majority of the cells progress through this sequence forming the intact individual, a few cells leave this pathway to become sequestered as reserve quiescent stem cells. Based on previous and current studies (Caplan 1991; Young et al. 1991, 1992a,b, 1993, 1995, 1998a,b, 1999, 2001a,b, 2003a–c; Lucas et al. 1993, 1995, 1996a; Pate et al. 1993; Rogers et al. 1995; Saito et al. 1995; Dixon et al. 1996; Warejcka et al. 1996; Clarke et al. 2000; Woodbury et al. 2000; McKinney-Freeman et al. 2002; Young and Black 2003), we propose at least 11 categories of quiescent reserve stem cells present within postnatal animals, including humans. The proposed categories are pluripotent epiblastic-like stem cells, pluripotent ectodermal stem cells, pluripotent epidermal stem cells, pluripotent neuronal stem cells, pluripotent neural crest stem cells, pluripotent mesenchymal (mesodermal) stem cells, pluripotent endodermal stem cells, multipotent progenitor stem cells, tripotent progenitor stem cells, bipotent progenitor stem cells, and unipotent progenitor stem cells. Thus, these quiescent reserve stem cells, i.e., lineage-committed progenitor stem cells and lineage-uncommitted pluripotent stem cells, would provide for the continual maintenance and repair of the postnatal organism after birth.

Acknowledgements. I would like to thank Dr. P.M. Johnston, emeritus Professor and Chairman, Dept. of Zoology, University of Arkansas, Fayetteville, AR. Without his three tenets for research, i.e., 'don't always believe negative consensus opinion, even if its in the scientific literature', 'know your model system', and 'tissue never lies', I would not have progressed as far with the adult stem cell research. I would like to thank my collaborators, co-authors and assistants for their insight and work ethic and John Knight for photographic assistance. Antibodies were obtained from the Developmental Studies Hybridoma Bank developed under the auspices of the NICHD and maintained by The University of Iowa, Department of Biological Sciences, Iowa City, IA 52242 and were developed by: MC480, MC631, and MC813-70, D. Solter; FORSE-1, P. Patterson; RAT-401, S. Hockfield; RT-97, J. Wood; 8A2, by V. Lemmon; Rip, S. Hockfield; SV2, K.M. Buckley; VM-1, V.B. Morhenn; 151-Ig, A. Hubbard; 40E-C, A. Alvarez-Buylla; F5D, W.E. Wright; MF-20, D.A. Fischman; ALD-58, D.A. Fischman; A4.74, H.M. Blau; CIIC1, R. Holmdahl and K. Rubin; D1-9, X.-J. Ye and K. Terato; 9/30/8A4, B. Caterson; 12/21/1C6, B. Caterson; 12C5, R.A. Asher; WV1D1, M. Solursh and A. Frazen;, MP111, M. Solursh and A. Frazen; P2B1, E.A. Wayner and G. Vercellotti; P8B1, E.A. Wayner and T. LeBien; P2H3, E.A. Wayner and G. Vercellotti. This research was supported by grants from Lucille M. and Henry O.Young Estate Trust, Rubye Ryle Smith Charitable Trust, MedCen Community Health Foundation, and MorphoGen Pharmaceuticals, Inc.

References

Ailhaud G, Grimaldi P, Negrel R (1992) Cellular and molecular aspects of adipose tissue development. Annu Rev Nutr 12:207–33

Beresford JN (1989) Osteogenic stem cells and the stromal system of bone and marrow. Clin Orthop 240:270–80

Bjornson CR, Rietze RL, Reynolds BA, Magli MC, Vescovi AL (1999) Turning brain into blood: a hematopoietic fate adopted by adult neural stem cells in vivo. Science 283:534–537

Campion, DR (1984) The muscle satellite cell: a review. Int Rev Cytol 87:225–251

Caplan AI (1991) Mesenchymal stem cells. J Orthop Res 9:641–650

Caplan AI, Elyaderani M, Mochizuki Y, Wakitani S, Goldberg V (1997) Principles of cartilage repair and regeneration. Clin Orthop Rel Res 342:254–269

Clarke DL, Johansson CB, Wilbertz J, Veress B, Nilsson E, Karlstrom H, Lendahl U, Frisen J (2000) Generalized potential of adult neural stem cells. Science 288:1660–1663

Cruess, RL (1982) The Musculoskeletal system: embryology, biochemistry, and physiology. New York, Churchill Livingston, pp. 1–33, 109–169, 255–287

Dixon K, Murphy RW, Southerland SS, Young HE, Dalton ML, Lucas PA (1996) Recombinant human bone morphogenetic proteins-2 and 4 (rhBMP-2 and rhBMP-4) induce several mesenchymal phenotypes in culture. Wound Rep Reg 4:374–380

Donovan PJ, Gearhart J (2001) The end of the beginning for pluripotent stem cells. Nature 414:92–927

Eglitis MA, Mezey E (1997) Hematopoietic cells differentiate into both microglia and macroglia in the brains of adult mice. Proc Natl Acad Sci USA 94:4080–4085

Farber J (1959) An experimental analysis of regional organization in the regenerating forelimb of the axolotl (*Ambystoma mexicanum*). Arch Biol 71:1–72

Grounds MD, Garrett KL, Lai MC, Wright WE, Beilharz MW (1992) Identification of muscle precursor cells in vivo by use of MyoD1 and myogenin probes. Cell Tiss Res 267:99–104

Hayflick L (1965) The limited *in vitro* lifetime of human diploid cell strains. Exper Cell Res 37:614–636

Iten LE, Bryant SV (1973) Forelimb regeneration from different levels of amputation in the newt. *Notophthalamus viridescens*. Length, rate, stage. Wilhelm Roux Archiv 173:77–89

Jackson KA, Mi T, Goodell MA (1999) Hematopoietic potential of stem cells isolated from murine skeletal muscle. Proc Natl Acad Sci U S A 96:14482–14486

Kishimoto T, Kikutani H, Borne AEGKrvd, Goyert SM, Mason D, Miyasaka M, Moretta L, Okumura K, Shaw S, Springer T, Sugamura K, Zola H (1997) Leucocyte Typing VI, White Cell Differentiation Antigens. Garland Publishing, Hamden, CT

Lucas PA, Calcutt AF, Ossi P, Young HE, Southerland SS (1993) Mesenchymal stem cells from granulation tissue. J Cell Biochem 17E:122

Lucas PA, Calcutt AF, Southerland SS, Wilson JA, Harvey RL, Warejcka D, Young HE (1995) A population of cells resident within embryonic and newborn rat skeletal muscle is capable of differentiating into multiple mesodermal phenotypes. Wound Rep Reg 3:457–468

Lucas PA, Warejcka DJ, Zhang L-M, Newman WH, Young HE (1996a) Effect of rat mesenchymal stem cells on the development of abdominal adhesions after surgery. J Surg Res 62:229–232

Lucas PA, Warejcka DJ, Young HE, Lee BY (1996b) Formation of abdominal adhesions is inhibited by antibodies to transforming growth factor-beta1. J Surg Res 65:135–138

Mauro A (1961) Satellite cell of skeletal muscle fibers. J Biophys Biochem Cytol 9:493–498

McGuire, WP (1998) High-dose chemotherapy and autologous bone marrow or stem cell reconstitution for solid tumors. Curr Probl Cancer 22:135–137

McKinney-Freeman SL, Jackson KA, Camargo FD, Ferrari G, Mavillio F, Goodell MA (2002) Muscle-derived hematopoietic stem cells are hematopoietic in origin. Proc Natl Acad Sci USA 99:1341–1346

Owen M (1988) Marrow stromal cells. J Cell Sci Suppl 10:63–76

Palis J, Segel GB (1998) Developmental biology of erythropoiesis. Blood Rev 12:1061–1064

Pate DW, Southerland SS, Grande DA, Young HE, Lucas PA (1993) Isolation and differentiation of mesenchymal stem cells from rabbit muscle. Surg Forum XLIV:587–589

Petersen BE, Bowen WC, Patrene KD, Mars WM, Sullivan AK, Murase N, Boggs SS, Greenberger JS, Goff JP (1999) Bone marrow as a potential source of hepatic oval cells. Science 284:1168–1170

Pittenger MF, Mackay AM, Beck SC, Jaiswal RK, Douglas R, Mosca JD, Moorman MA, Simonetti DW, Craig S, Marshak DR (1999) Multilineage potential of adult human mesenchymal stem cells. Science 148:143–147

Pritchette WH, Dent JN (1972) The role of size in the rate of limb regeneration in the adult newt. Growth 36:275–289

Prockop DJ (1997) Marrow stromal cells as stem cells for nonhematopoietic tissues. Science 276:71–74

Ratajczak MZ, Pletcher CH, Marlicz W, Machlinski B, Moore J, Wasik M, Ratajczak J, Gewirtz AM (1998) CD34+, kit+, rhodamine 123 (low) phenotype identifies a marrow cell population highly enriched for human hematopoietic stem cells. Leukemia 12:942–950

Rogers JJ, Young HE, Adkison LR, Lucas PA, Black AC Jr (1995) Differentiation factors induce expression of muscle, fat, cartilage, and bone in a clone of mouse pluripotent mesenchymal stem cells. Amer Surg 61:231–236

Saito T, Dennis JE, Lennon DP, Young RG, Caplan AI (1995) Myogenic expression of mesenchymal stem cells within myotubes of *mdx* mice *in vitro* and *in vivo*. Tiss Eng 1:327–343

Scadding SR (1977) Phylogenetic distribution of limb regeneration capacity in adult amphibia. J Exp Zool 202:57–68

Shamblott MJ, Axelman J, Wang S, Bugg EM, Littlefield JW, Donovan PJ, Blumenthal PD, Huggins GR, Gearhart JD (1998) Derivation of pluripotent stem cells from cultured human primordial germ cells. Proc Natl Acad Sci USA 95:13726–13731

Singer M (1951) Induction of regeneration of the forelimb of the frog by augmentation of the nerve supply. Proc Soc Exp Biol Med 76:413–416

Singer M (1978) On the nature of the neurotrophic phenomenon in urodele regeneration. Am Zool 18:829–841

Stocum DL (1979) Stages of forelimb regeneration in *Ambystoma maculatum*. J ExP Zool 209:395–416

Stocum DL (1998) Regenerative biology and engineering: strategies for tissue restoration. Wound Rep Reg 6:276–290

Tank PW, Carlson BM, Connelly TG (1976) A staging system for forelimb regeneration in the axolotl, *Ambystoma mexicanum*. J Morph 150:117–128

Tank PW, Holder N (1981) Pattern regulation in the regenerating limbs of urodele amphibians. Quart Rev Biol 56:113–142.

Thomson JA, Itskovitz-Eldor J, Shapiro SS, Waknitz MA, Swiergiel JJ, Marshall VS, Jones JM (1998) Embryonic stem cell lines derived from human blastocysts. Science 282:1145–1147

Thornton CS (1968) Amphibian limb regeneration. In: Brachet J, King TJV (eds) Advances in Morphogenesis, Academic Press, New York, Vol. 7, pp. 205–249

Toole BP, Gross J (1971) The extracellular matrix of the regenerating newt limb: synthesis and removal of hyaluronate prior to differentiation. Dev Biol 25:57–77

Vierck JL, McNamara JP, Dodson MV (1996) Proliferation and differentiation of progeny of ovine unilocular fat cells (adipofibroblasts). In Vitro Cell Dev Biol Anim 32:564–572

Vescovi AL, Galli R, Gritti A (2001) The neural stem cells and their transdifferentiation capacity. Biomed Pharmacother 55:201–205

Warejcka DJ, Harvey R, Taylor BJ, Young HE, Lucas PA (1996) A population of cells isolated from rat heart capable of differentiating into several mesodermal phenotypes. J Surg Res 62:233–242

Woodbury D, Schwarz EJ, Prockop DJ, Black IB (2000) Adult rat and human bone marrow stromal cells differentiate into neurons. J Neurosci Res 61:364–370

Young HE (1977a) Epidermal ridge formation during limb regeneration in the adult salamander, *Ambystoma annulatum*. Proc Ark Acad Sci 31:107–109

Young HE (1977b) Limb regeneration in the adult salamander, *Ambystoma annulatum* Cope 1889 (Amphibia: Ambystomatidae). Univ. Arkansas Library Press, Fayetteville

Young HE (1977c) Anomalies during limb regeneration in the adult salamander, *Ambystoma annulatum*. Proc Ark Acad Sci 31:110–111

Young HE (1983) A Temporal Examination of Glycoconjugates During the Initiation Phase of Limb Regeneration in Adult Ambystoma. Texas Tech Univ. Library Press, Lubbock

Young HE (2000) Stem cells and tissue engineering. In: Huard J, Fu FH (eds) Gene Therapy in Orthopaedic and Sports Medicine, Springer-Verlag New York, Inc., New York, NY, pp. 143–173

Young HE, Bailey CF, Dalley BK (1983a) Environmental conditions prerequisite for complete limb regeneration in the postmetamorphic adult land-phase salamander, Ambystoma. Anat Rec 206:289–294

Young HE, Bailey CF, Dalley BK (1983b) Gross morphological analysis of limb regeneration in postmetamorphic adult Ambystoma. Anat Rec 206:295–306

Young HE, Dalley BK, Markwald RR (1983c) Identification of hyaluronate within peripheral nervous tissue matrices during limb regeneration. In: Coates PW, Markwald RR, Kenny AD (eds), Developing and Regenerating Vertebrate Nervous Systems, Neurology and Neurobiology, Alan R. Liss, Inc., New York, 6:175–183

Young HE, Bailey CF, Markwald RR, Dalley BK (1985) Histological analysis of limb regeneration in postmetamorphic adult Ambystoma. Anat Rec 212:183–194

Young HE, Dalley BK, Markwald RR (1989a) Glycoconjugates in normal wound tissue matrices during the initiation phase of limb regeneration in adult Ambystoma. Anat Rec 223:223–230

Young HE, Dalley BK, Markwald RR (1989b) Effect of selected denervations on glycoconjugate composition and tissue morphology during the initiation phase of limb regeneration in adult Ambystoma. Anat Rec 223:231–241

Young HE, Morrison DC, Martin JD, Lucas PA (1991) Cryopreservation of embryonic chick myogenic lineage-committed stem cells. J Tiss Cult Meth 13:275–284

Young HE, Ceballos EM, Smith JC, Lucas PA, Morrison DC (1992a) Isolation of embryonic chick myosatellite and pluripotent mesenchymal stem cells. J Tiss Cult Meth 14:85–92

Young HE, Sippel J, Putnam LS, Lucas PA, Morrison DC (1992b) Enzyme-linked immuno-culture assay. J Tiss Cult Meth 14:31–36

Young HE, Ceballos EM, Smith JC, Mancini ML, Wright RP, Ragan BL, Bushell I, Lucas PA (1993) Pluripotent mesenchymal stem cells reside within avian connective tissue matrices. In Vitro Cell Dev Biol Anim 29A:723–736

Young HE, Mancini ML, Wright RP, Smith JC, Black AC Jr, Reagan CR, Lucas PA (1995) Mesenchymal stem cells reside within the connective tissues of many organs. Dev Dynam 202:137–144

Young HE, Wright RP, Mancini ML, Lucas PA, Reagan CR, Black AC Jr (1998a) Bioactive factors affect proliferation and phenotypic expression in pluripotent and progenitor mesenchymal stem cells. Wound Rep Reg 6:65–75

Young HE, Rogers JJ, Adkison LR, Lucas PA, Black AC Jr (1998b) Muscle morphogenetic protein induces myogenic gene expression in Swiss-3T3 cells. Wound Rep Reg 6:530–541

Young HE, Steele TA, Bray RA, Detmer K, Blake LW, Lucas PA, Black AC Jr (1999) Human pluripotent and progenitor cells display cell surface cluster differentiation markers CD10, CD13, CD56, and MHC Class-I. Proc Soc Exp Biol Med 221:63–71

Young HE, Duplaa C, Young TM, Floyd JA, Reeves ML, Davis KH, Mancini GJ, Eaton ME, Hill JD, Thomas K, Austin T, Edwards C, Cuzzourt J, Parikh A, Groom J, Hudson J, Black AC Jr. (2001a) Clonogenic analysis reveals reserve stem cells in postnatal mammals. I. Pluripotent mesenchymal stem cells. Anat Rec 263:350–360

Young HE, Steele T, Bray RA, Hudson J, Floyd JA, Hawkins K, Thomas K, Austin T, Edwards C, Cuzzourt J, Duenzl M, Lucas PA, Black AC Jr. (2001b) Human reserve pluripotent mesenchymal stem cells are present in the connective tissues of skeletal muscle and dermis derived from fetal, adult, and geriatric donors. Anat Rec 264:51–62

Young HE, Black Jr AC (2003) Adult stem cells. Anat Rec (in press)

Young HE, Duplaa C, Romero-Ramos M, Chesselet M-F, Vourc'h P, Yost MJ, Ericson K, Terracio L, Asahara T, Masuda H, Tamura-Ninomiya S, Detmer K, Bray RA, Steele TA, Hixson D, El-Kalay M, Tobin BW, Russ RD, Horst MN, Floyd JA, Henson NL, Hawkins KC, Groom J, Blake L, Bland LJ, Thompson AJ, Kirincich A, Moreau C, Hudson J, Bowyer III FP, Lin TJ, Black Jr AC (2003a) Tissue engineering using adult reserve pluripotent stem cells. J Cell Biochem Biophys (in press)

Young HE, Duplaa C, Yost MJ, Henson NL, Floyd JA, Detmer K, Thompson AJ, Powell SW, Gamblin TC, Kizziah K, Holland BJ, Boev A, Van de Water JM, Godbee DC, Edwards CR, Wu E, Cawley C, Edwards PD, Macgregor A, Bozof R, Thompson TM, Petro Jr GJ, Shelton HM, McCampbell BL, Mills JC, Flynt FL, Steele TA,

Kearney M, Kirincich-Greathead A, Hardy W, Young PR, Amin AV, Williams RS, Horton MM, McGuinn S, Ericson K, Terracio L, Moreau C, Hixson D, Tobin BW, Hudson J, Bowyer III FP, Black Jr AC. (2003b) Clonogenic analysis reveals reserve stem cells in postnatal mammals. II. Pluripotent epiblastic-like stem cells. Anat Rec (submitted)

Young HE, Bray RA, Detmer K, Yoder MC, Henson NL, Floyd JA, Hawkins KC, Groom J, Duenzl M, Thompson AJ, Hixson D, Hudson J, Bowyer III FP, Lin TJ, Black AC Jr. (2003c) Reserve pluripotent stem cells resembling embryonic stem cells are present in adult humans. Anat Rec (submitted)

Genetic Regulation of Mouse Stem Cells: Identification of Two Keratinocyte Stem Cell Regulatory Loci

N. V. Popova · R. J. Morris

Department of Dermatology, Columbia University College of Physicians and Surgeons, 630 West 168th Street, VC-15, New York, NY 10032, USA

E-mail: nvp2002@ columbia.edu

1 Introduction 112

2 Methods 114
2.1 Mice 114
2.2 Histology 115
2.3 Phenotyping 115
2.4 Statistical Analysis 115
2.5 Genotype Analysis 116

3 Results 116
3.1 Mouse KC Number Is a Quantitative Multigenic Trait 116
3.2 Mouse Keratinocyte Colony Size Is Genetically Regulated 123
3.3 Identification of Keratinocyte Stem Cell Regulatory Loci 123
3.4 Keratinocyte Stem Cells and Two-Stage Skin Carcinogenesis 131

4 Conclusion 134

References 135

Abstract It is well documented that the bulge of hair follicle is a 'niche' for a significant population of mouse keratinocyte stem cells, and 95% of rodent clonogenic keratinocytes originate from the bulge region. The ability to form colonies in vitro is a well recognized test for keratinocyte stem cells. We analyzed the epidermis of seven mouse strains and their segregating crosses [(BALB/c×C57BL/6)F1; (BALB/c×CB6F1); (C57BL/6×CB6F1); (CBF1×CBF1)F2] for their clonogenic activity in vitro. We found that keratinocyte colony (KC) number is a new quantitative multigenic trait. The analysis of KC size in two parental strains (C57BL/6 and BALB/c), the F1 generation and the segregating crosses demonstrated that the size of KC is a quantitative complex trait also. We determined that mouse epidermis has at least two subpopulations of keratinocytes that gave small (<2 mm^2) and large (>2 mm^2) colonies. The differences in the number of small and large colonies between parental strains (C57BL/6, BALB/c) were significant ($P<0.01$). A genome-wide scan of the intercross and the two backcrosses maps the number of small KC to the central region of mouse Chromosome 9 (genomewide P value=0.01). We define this locus as Ksc1. The proximal region of chromosome 4 is associated with the high number of large KC. We defined this locus as Ksc2. We found that Ksc1 and minor loci on chromosomes 6 and 7 map close, if not equal to, loci associated with mouse skin carcinogenesis. We conclude that mouse epidermis has at least two subpopulations of clonogenic keratinocyte stem cells that are regulated by different genes. We suggest that keratinocyte stem cells responsible for small colonies may play a major role in the regulation of resistance or sensitivity to skin carcinogenesis. Investigation of the genes regulating the stem cell number should provide new insight into the mechanisms of skin carcinogenesis, and should help to develop new approaches for therapies not only against active proliferating tumor cells but also quiescent tumor stem cells.

1 Introduction

The epidermis is a continuously renewing tissue in which keratinocytes are organized into multiple layers and undergo terminal differentiation (stratum corneum) as they move through the suprabasal layers toward the skin surface. Stem cells, which are responsible for skin homeostasis and are required throughout life, enable the maintenance of the self-renewing epidermis.

Stem cells produce daughter stem cells and nonstem cells. The nonstem cell progenitors, known as transit-amplifying cells, have limited proliferative capacity and are identified as a pool of rapidly proliferating cells that are lost from the basal layer to terminal differentiation within 4–5 days (Potten 1986; Morris et al. 1985). Keratinocyte stem cells represent a small population of undifferentiated slow-cycling cells with self-renewal capacity and extensive proliferative and differentiation potential in vivo and in vitro. These stem cells are generally thought to reside in the bulge region of the hair follicle (Cotsarelis et al. 1999; Morris and Potten 1999; Watt 2001; Fuchs and Raghavan 2002) (Fig. 1).

Two traditional approaches have been used to identify epidermal stem cells. One is the identification of label-retaining cells remaining in the

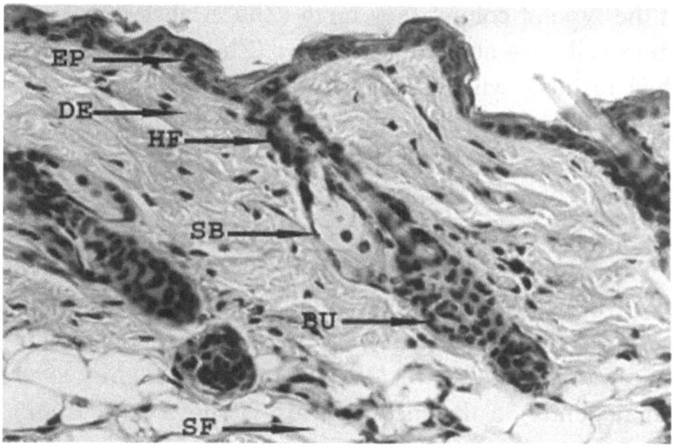

Fig. 1 Photograph of mouse skin: *EP*, epidermis; *DE*, dermis; *HF*, hair follicle, catagen stage; *SB*, sebaceous gland; *BU*, bulge region; *SF*, subcutaneous fat. Original magnification, ×400

hair follicle bulge region of adult mice injected with H^3-thymidine or BrdU as neonates. The second approach involves analysis of the proliferation of epidermal keratinocytes cultured at clonal density and assessment of serial passage number. Therefore the multipotential stem cells share an ability for label-retention and clone-formation.

The experiments of Barrandon and colleagues (Kobayashi et al. 1993; Rochat et al. 1994; Oshima et al. 2001) and Morris (Morris and Potten 1994) demonstrated that slow-cycling label-retaining multipotential keratinocyte stem cells in the hair follicle bulge region have clonogenic activity. Several lines of evidence suggest that keratinocyte colony (KC)-forming cells are closely related, or identical to multipotential stem cells. The bulge region has low mitotic and DNA synthetic activity in vivo (Oshima et al. 2001), and most of the visible KCs originate from this region (Kobayashi et al. 1993; Rochat et al. 1994; Oshima et al. 2001). In contrast, the highly mitotic matrix region of the hair follicle has extremely low clonogenic activity in vitro (Oshima et al. 2001). In addition, slow-cycling (label-retaining) epidermal cells in vivo behave like clonogenic stem cells in vitro (Morris and Potten 1994). Thus, most or all visible KCs are thought to originate from stem cells.

Fiona Watt and co-workers suggest that human keratinocytes with characteristics of stem cells can be distinguished from transit-amplifying cells based on the type of colony they form (Zhu et al. 1999). Colonies formed by stem cells are able to self-renew (Zhu et al. 1999; R.J. Morris, unpublished results), whereas transit-amplifying colonies contain fewer than 30–40 cells, all of which undergo terminal differentiation (Zhu et al. 1999), and can be identified by microscopy at low magnification. Thus, most visible colonies appear to originate from keratinocyte stem cells.

We have identified several factors that might influence the number of KCs. Although colony number remains virtually constant throughout most of adult life in mice (Morris et al. 1988), the number increases during tumor promotion of initiated skin (Morris et al. 1988). Moreover, the colony number depends on the mouse strain (Popova et al. 2002). Thus, amplification of keratinocyte stem cells may play a role in skin carcinogenesis, and KC number might be genetically regulated. The identification of regulatory genes involved in stem cell amplification would enhance the understanding of normal skin homeostasis and skin carcinogenesis. To address this issue, we carried out genetic analyses of KC formation in different mouse strains.

2 Methods

2.1 Mice

Six-week-old C57BL/6 [C57BL/6NCrIBR], DBA/2 [DBA/2NCrIBR], BALB/c [BALB/cAnNCrIBR], CD-1 [Crl:CD-1 (ICR)], C3H [C3H/HeN-CrIBR], FVB [FVB/NCrIBR], (BALB/c×C57BL/6)F1, C57BL/6×(BALB/c×C57BL/6)F1 and BALB/c×(BALB/c×C57BL/6)F1 backcross, (BALB/c×C57BL/6)F2 female mice were obtained from Charles River Laboratories (Wilmington, Massachusetts; www.criver.com). Six-week-old SENCAR female mice were obtained from NCI Frederick Cancer Research Center (Bethesda, MD). Groups of each mouse strains ranged between 24 and 57 animals; 30 F1, 44–45 backcrosses, and 104 F2 mice were used. A total of 496 mice were analyzed. All experiments were carried out with mice at 7 weeks of age, when skin was still pink, indicating the resting stage of the hair follicle growth cycle.

2.2 Histology

Skin of mice was fixed in 10% neutral buffered formalin, embedded in paraffin and prepared by serial sections (5 μm). Slides were stained with hematoxylin and eosin.

2.3 Phenotyping

Epidermal cells including those from hair follicles were harvested from the dorsal skin and grown from single cells on irradiated 3T3 (Swiss) mouse fibroblasts (American Type Culture Collection, Rockville, MD) for intervals of 2 and 4 weeks (Popova et al. 2002, 2003). Keratinocytes harvested (Morris 1994) from each mouse were placed into four 60-mm plastic dishes (1,000 viable cells/dish), fixed with neutral buffered formalin, and stained with Rhodamine B. KC number and colony size were analyzed with a Fluor-S MultiImager (BIO-RAD) after 2 weeks of cultivation (Popova et al. 2002, 2003). Colony size ranged from 0.1 to 12 mm^2. Average KC number in a 0.1–2 mm^2 area in duplicate dishes was designated as a 'small' colony phenotype, and average KC number in a > 2 mm^2 area was considered a 'large' colony phenotype (Figs. 1 and 2). Finally, we had two phenotypes. The first of our phenotypes was the number of small colonies from each individual animal of each segregating crosses, the second was the number of large colonies.

2.4 Statistical Analysis

A balanced incomplete block design experiment (Box et al. 1978) was carried out with 12 replicates from each mouse strain and for the 6F1 mice. Differences in KC numbers between different mouse groups were analyzed using the Wilcoxon rank sum test and extension of the two-way Anova.

2.5 Genotype Analysis

DNA was prepared from mouse liver by a standard proteinase K procedure (Wolff and Gemmill 1997). Genotyping was performed by polymerase chain reaction (PCR) as described previously (Angel et al. 2000), using microsatellite primers (Research Genetics, Hunstsville, AL). Microsatellite markers (a total of 114) were chosen to cover all known chromosomal segments of sequence-tagged sites originating in autosomes from C57BL/6 and BALB/c mice. The maximum distance between two adjacent markers was no more than 15 cM. PCR products were analyzed by electrophoresis on 3% agarose gels, stained with ethidium bromide and visualized under ultraviolet light. Statistical analyses were performed using Qlink software kindly provided by N. Drinkwater (University of Wisconsin). Loci were ordered using the map locations reported in the 2000 Mouse Chromosome Committee Reports obtained from the Mouse Genome Database (http://www.informatics.jax.org/ccr/searches/index.cgi?year=2000).

3 Results

3.1 Mouse KC Number Is a Quantitative Multigenic Trait

These studies arose from our initial findings that five inbred (C57BL/6, C3H, DBA/2, BALB/c, FVB) and two outbred (CD-1, SENCAR) mouse strains significantly differ in the number of KC-forming cells per 1000 viable cells isolated from skin taken from the back of a mouse (Fig. 2).

The KC number was not associated in any obvious way with parameters such as numbers of hair follicles, basal or superbasal cells, mitotic cells/cm^2 of skin (Table 1), or percentage of viable cells/cm^2 of skin. Furthermore, there was no significant difference between high and low KC-expressing mice in the number of attached cells. This finding suggested that the different KC numbers reflected differences in the different genetic background of these mouse strains.

To avoid potential bias in KC number associated with day-to-day variation, we used a balanced incomplete block design (Box et al. 1978) for the experiments. Twelve replicates from each strain were analyzed, and

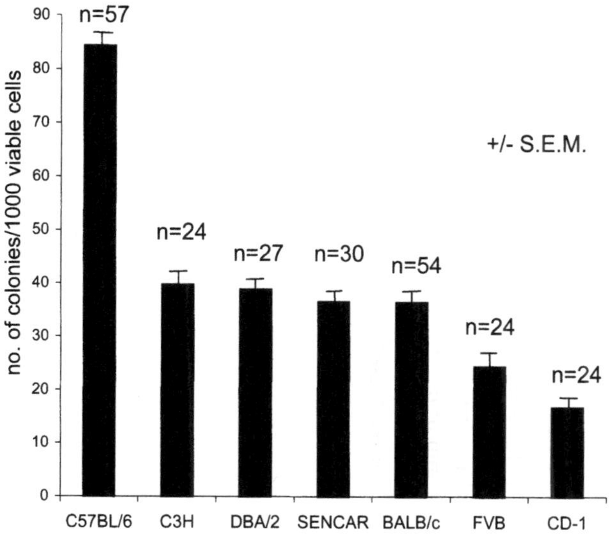

Fig. 2 Keratinocyte colony-forming ability in seven mouse strains (mean±SEM/1000 viable cells). C57BL/6, 84.3±3.2; C3H, 39.7±4.5; DBA/2, 36.8±2.1; SENCAR, 36.5±2.7; BALB/c, 36.4±1.7; FVB, 24.4±2.9, and CD-1, 16.8±1.5

each replicate of every strain was compared with every other strain the same number of times. Each block included less than the total number of strains in the full experiment. The same number of strains was studied in a random order in each experiment and, potential environmental influences on phenotypes were minimized.

KC numbers were highly stable and reproducible for each of the mouse strains. Mean counts (±SEM) per 1,000 viable cells ranged from 84.3±3.2 for C57BL/6, to 16.8±1.5 for CD-1 (Fig. 2) and a subset of these major groups with respect to KC numbers was identified: C57BL/6>>SENCAR=C3H=DBA/2=BALB/c>FVB=CD-1. KC numbers were significantly different between these three groups of mice. The mouse strains from each subset have different genetic origins: based on the genealogies of mouse inbred strains (Beck et al. 2000), C57BL/6 mice are in category E; BALB/c, C3H, DBA/2 mice are in category B; and CD-1 and FVB mice are in category A (Fig. 3). The finding of significant differences in KC numbers in three subsets of inbred, genetically distinct mouse strains further suggests that KC number is genetically regulated. Moreover, F1 progeny of the genetically distinct C57BL/6 and BALB/c inbred strains, which differ significantly ($P<0.01$) in KC number (Table 2;

Table 1 Similar cutaneous parameters in different strains of mice

Strain	Basal nuclei per mm IFE[a]	Suprabasal nuclei per mm IFE[a]	Total nuclei per mm IFE[a]	Nucleated cell layers[b]	Mitoses per mm IFE[c]	Hair follicles per cm^2 of skin[d]
C57Bl/6	154.7±2.338	55.8±3.371	210.5±3.886	2	2.67±1.51	2435±102.4
SENCAR	154.5±1.871	57.5±2.168	212.0±1.549	2	2.17±1.29	2088±148.4
DBA/2	150.8±1.472	58.2±1.47	209.0±2.366	2	3.33±1.33	2166±126.8
BALB/c	151.8±2.563	56.7±3.50	208.5±3.728	2	2.75±0.987	2290±126.8
CD-1	152.8±1.169	55.7±3.72	208.5±4.037	2	2.67±1.33	2339±184.7

[a] Nuclei were counted in 1 mm of interfollicular epidermis with a Zeiss microscope equipped with a ×100 planapochromatic objective. Mean±SD, n=6 mice.
[b] The epidermis of all strains consisted of a single basal layer covered by a single suprabasal layer.
[c] Colchicine was administered 3 h prior to sacrifice. Mitotic figures were counted in 2 mm of interfollicular epidermis and were expressed as the number per mm. Mean±SD, n=6 mice.
[d] Hair follicles were counted per cm^2 of interfollicular epidermis with a ×10 objective and expressed as the number per cm^2 of skin. Mean±SD, n=6 mice.

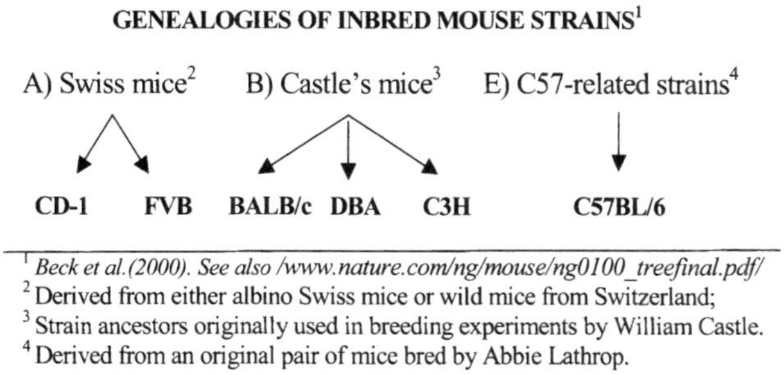

Fig. 3 Genealogies of C57BL/6, BALB/c, DBA, C3H, CD-1, FVB mouse strains. (From Beck et al. 2000)

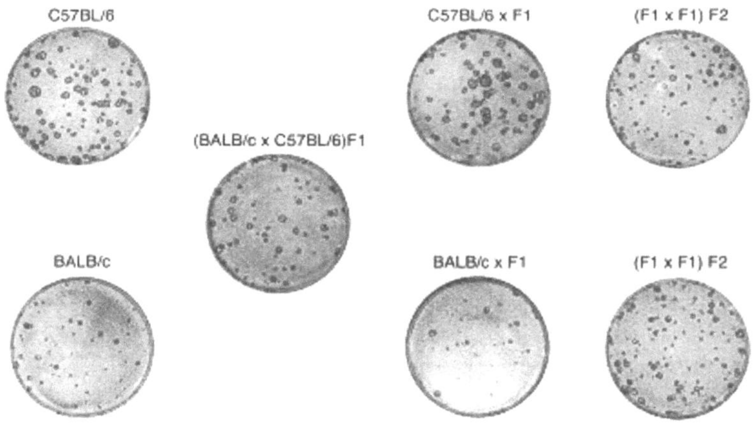

Fig. 4 Keratinocyte colony formation after 2 weeks of cultivation from parental strains C57BL/6 and BALB/c and their segregating genetic crosses (BALB/c×C57BL/6) F1, C57BL/5× F1, BALB/c×F1, (F1×F1) F2

Fig. 4) displayed intermediate KC numbers. The difference between the (C57BL/6×F1) and the (BALB/c×F1) backcrosses KC number was highly significant ($P<0.001$) respectively (Table 2; Fig. 4). Consequently the KC number is genetically regulated.

When the KC number per mouse in the (C57BL/6×F1) was plotted against that in the (BALB/c×F1) backcross (Fig. 5), curves did not overlap. At least seven mice with lowest number of colonies in the BALB/c×F1 backcross were required to reach the lowest KC number in the (C57BL/6×F1) backcross. At least 43 (BALB/c×F1) mice were required to

Table 2 Keratinocyte colony number

Mice	No. of mice	No. of colonies	
C57BL/6	57	84.3±24.2[a]	83.0 (50%)[b]
BALB/c	54	36.4±12.2[a]	34.2 (34.6%)[b]
CB6F1	30	53.0±22.1[a]	49.3 (47%)[b]
C57Bl/6×CB6F1	45	74.2±27.2[a]	72.2 (41%)[b]
BALB/c×CB6F1	44	42.9±16.3[a]	39.5 (52%)[b]
CB6F2	104	65.5±26[a]	60.7 (56%)[b]

[a] Mean KC number±SD.

[b] Values represent the mean KC numbed estimated using $\log_{10}$ of the observations where the anti-log of the average is reported as the mean. Standard deviations are reported as percent (within parentheses) because in the log distribution the SD is proportional to the observation.

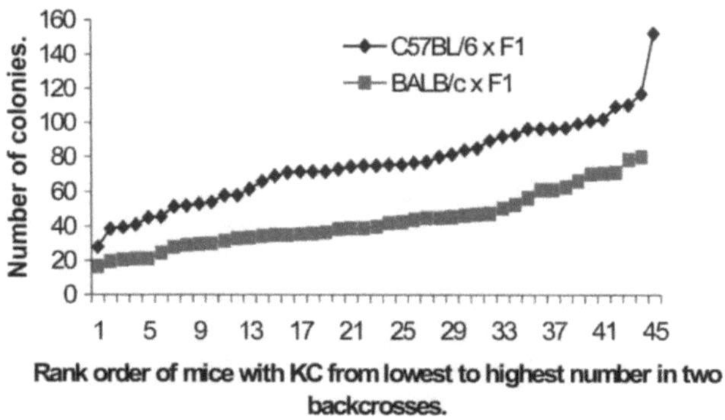

Fig. 5 Number of colonies per mouse in each of two backcrosses (C57BL/6×F1, n= 45; BALB/c×F1, n= 44) plotted from lowest to highest number. (Wilcoxon rank sum test)

reach the median in the (C57Bl/6×F1) backcross determined by Wilcoxon rank sum test. These data demonstrate the segregation of genes (or alleles) associated with high and low KC number in these two backcrosses.

Analysis according to a modification (Green 1981) of Mendelian one-locus (dominant recessive, codominant or partially dominant, semidominant, recessive lethal) genotypic segregation ratios in the progeny of intercross and backcrosses revealed no clear association of KC number with a single-locus autosomal model.

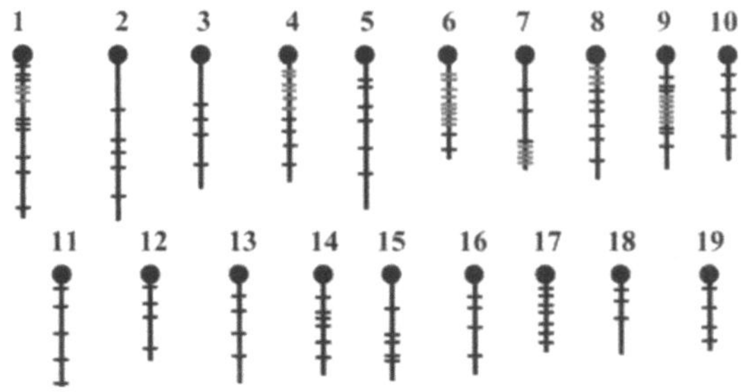

Fig. 6 Chromosomal map location of mouse microsatellite markers used in association with total KC number in the intercross of C57BL/6 and BALB/c mice

Several models were evaluated for the genetic determination of quantitative traits as an indicator of the number of genes controlling KC number. Nonparametric distributions were transformed into the distributions close to normal by estimating mean KC numbers as anti-log values (Table 2). Using the method of Elston and Steward (1973), we evaluated models of: two linked loci with equal and additive effect; two linked loci with equal and additive genes at each locus; and equal and additive effect for two loci or for all four genes. Our result indicated that KC number is regulated by two (or more) loci with additive but not necessarily equal effect, and is therefore a quantitative multigenic trait.

Linkage analysis of two backcrosses and the intercross of parental BALB/c and C57BL/6 mice revealed several loci in the proximal region of mouse chromosomes 1, 4, and 8, in the proximal and central region of chromosome 6, in the distal region of chromosome 7, and in the central region of chromosome 9 associated with total KC number (P=0.05–0.002) (Fig. 6). Because both the density of our microsatellite markers (15 cM intervals when available) and the number of mice per group (104 mice in the F2 generation) was sufficient for genomewide significance (Lander and Kruglyak 1995; McBrearty et al. 1998; Angel et al. 2000), we conclude that individual dishes contained a mixture of several subpopulations of clonogenic keratinocytes.

Table 3 Keratinocyte colony size in parental mouse strains C57BL/6 and BALB/c mice and their genetic crosses

Groups of mice[a]	Number of mice	Average colony size (mm^2)	Average area of colonies/dish	Average number of small colonies (%)[b]	Average number of large colonies (%)[c]	Maximal area of individual colonies (mm^2)
C57BL/6	7	2.3	191.9	54	46	12.56
BALB/c	12	1.0	52.8	91	9	4.9
CB6F1	10	1.6	102.3	71	29	9.9
C57BL/6×CB6F1	45			49	51	
BALB/c×CB6F1	44			73	27	
F2	104			83.8	16	

[a] Keratinocyte colonies were measured in 12 dishes of C57BL/6, in19 dishes of F1 hybrid, and in 21 dishes of BALB/c with Fluor-S™ MultiImager (BIO-RAD).
[b] Less than 2 mm^2.
[c] More than 2 mm^2.

3.2 Mouse Keratinocyte Colony Size Is Genetically Regulated

Analysis of keratinocyte colony size originating from parental C57Bl/6 and BALB/c mice and their genetic crosses (total of 242 mice analyzed) indicated a mean size of 2.3 mm^2 of 988 colonies from C57BL/6, while that from BALB/c mice was 1.0 mm^2 (mean size of 1,323 colonies) (Table 3). (BALB/c×C57BL/6) F1 mice had an intermediate colony size of 1.6 mm^2 (mean size of 1,236 colonies). The mean total area of all colonies/dish was higher in C57BL/6 mice (191.9 mm^2) then in BALB/c mice (52.5 mm^2). The F1 generation demonstrated an intermediate colony size (102.3 mm^2) (Table 3). The largest colony from C57BL/6 mice (12.56 mm^2) was 2.5 times greater than that from BALB/c mice (4.9 mm^2), while the colony size of the F1 progeny was intermediate (9.9 mm^2). These results suggest that KC size is genetically regulated and that clonogenic keratinocyte populations differ in C57BL/6 and BALB/c mice.

Analysis of the distribution frequency of different sized colonies from the parental strains and the F1 hybrid (Fig. 7) identified at least two subpopulations of clonogenic keratinocytes. The number of small colonies (0.1–2 mm^2) was significantly higher in BALB/c mice, while the number of large colonies (2.1–12 mm^2) was significantly higher in C57BL/6 mice ($P<0.0001$). The (BALB/c×C57BL/6) F1 showed an intermediate colony phenotype.

Together our data indicate that large and small KCs have different modes of inheritance in C57BL/6 and BALB/c mice suggesting that different genes regulate KC number.

3.3 Identification of Keratinocyte Stem Cell Regulatory Loci

In an effort to map genes regulating numbers of both small and large colonies, we separated all large and small colonies from each individual dish (Fig. 8) and monitored loci associated with numbers of small and large colonies in the two backcross and intercross mice using essentially the procedure described by Angel et al. in 2000. Clonogenic activity of keratinocytes from 104 intercross mice was initially investigated (Popova et al. 2003) and the same protocol was used to determine the number of small and large colonies in intercross and backcross mice. Use of 141

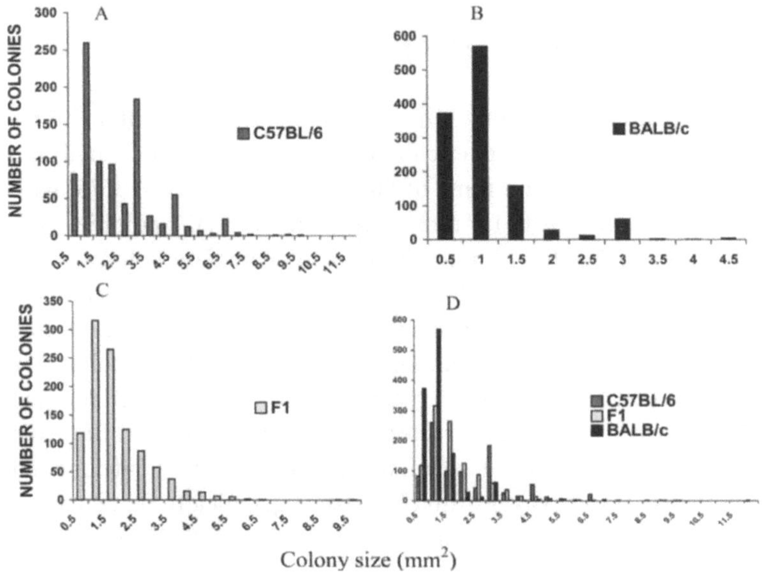

Fig. 7A–D Frequency distribution of keratinocyte colonies of different size (mm^2) in C57BL/6, BALB/c and (C57BL/6×BALB/c)F1 female mice after 2 weeks of cultivation. **A** 988 colonies from C57BL/6 mice. **B** 1,236 colonies from BALB/c mice. **C** 1,236 colonies from (BALB/c×C57BL/6) F1 mice. **D** Colonies from C57BL/6, BALB/c and F1 mice. Keratinocyte colony size was measured with a Fluor-S MultiImager (BIO-RAD)

microsatellite polymorphic markers allowed us to generate a complete genomewide scan at about 15-cM intervals. In areas in which there was a suggestion of linkage for small and large colony number, the density of markers was increased to obtain more accurate genetic dissection. Overall, genomic coverage was about 95.3% across the 19 autosomal chromosomes. The KC number in mice homozygous and heterozygous for each locus was compared using the Qlink program based on the Wilcoxon rank test. Tables 4, 5 and 6 list the microsatellite markers positively linked to the number of small or large colonies in the intercross or backcross.

Loci on chromosomes 1, 4, 6, 7, 8, 9, that were linked to total KC number were retained in the analysis of the number of small and large KCs. In particular, loci on chromosomes 1, 6, 7, 8, 9 associated with a number of small colonies (Tables 4 and 5), while a locus on chromosome 4 was

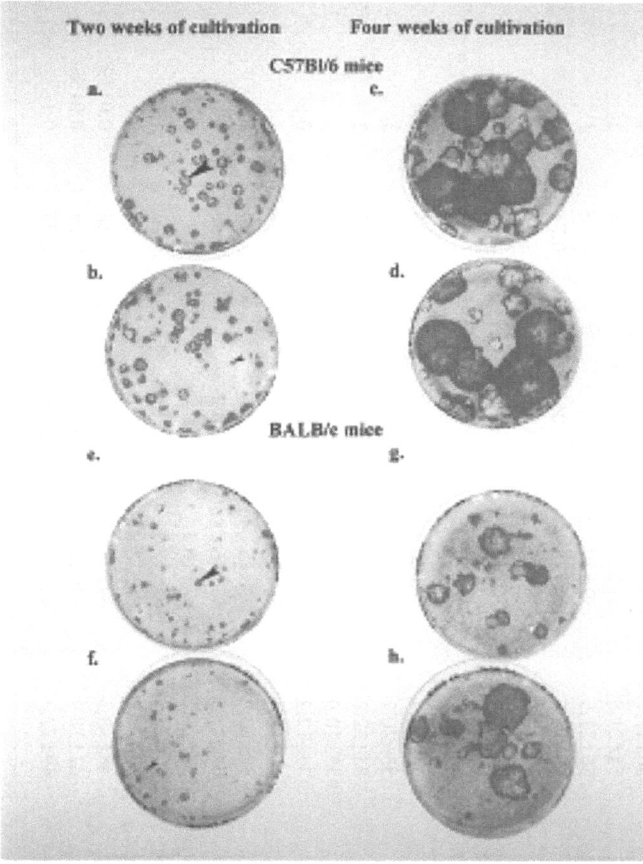

Fig. 8a–h Mouse keratinocyte colonies from C57BL/6 and BALB/c female mice after 2 and 4 weeks of cultivation on irradiated 3T3 (Swiss) mouse fibroblasts as a feeder layer. *Arrows* indicate large colonies from C57BL/6 and BALB/c (**a** and **e** respectively) mice after 2 weeks of cultivation and small colonies from C57BL/6 and BALB/c (**b** and **f**, respectively) mice after 2 weeks of cultivation

linked with number of large colonies (Table 1). Moreover, the level of significance for most of these loci increased upon reanalysis of the phenotype. Thus mouse skin has two major subpopulations of clonogenic keratinocytes regulated by different genes.

Multilocus analysis of chromosome 9 suggested that the most likely interval surrounding D9Mit262 (41 cM) and D9Mit236 (43 cM) in the

Table 4 Loci associated with small keratinocyte colony number in intercross (F2) between BALB/c and C57BL/6 mice

Marker[a]	Map location (cM)[b]	Colony number			*P* value[g]	LOD[h] score	Genome wide *P* value[i]
		aa[c]	*a*/*b*[d,e]	*bb*[f]			
D1Mit178	35	43±16.5	58±23.0	59±18.8	0.0058*	1.65	1
D1Mit132	43	47±19.5	56±23.1	61±18.6	0.01	1.42	1
D1Mit187	62	54±25.1	50±17.7	64±20.1	0.06	0.74	1
D6Mit86	0.5	50±20.9	54±21.8	61±21.0	0.04	0.88	1
D6Mit159	7	48±21	56±21.5	59±20.7	0.06	0.76	1
D6Mit74	20	45±20.0	58±18.5	57±25.4	0.06	0.74	1
D6Mit93	26	45±20.4	57±18.0	58±26.0	0.009*	1.48	1
D6Mit384	28	42±17.9	57±18.3	58±25.9	0.001*	2.10	1
D6Mit188	33	45±19.1	57±18.2	58±27.1	0.006*	1.59	1
D6Mit245	33	45±18.9	57±18.3	58±26.7	0.008*	1.49	1
D6Mit99	36	42±16.9	58±18.4	58±26.7	0.001*	2.25	0.4
D6Mit323	37	45±18.1	55±20.2	62±24.3	0.01*	1.25	1
D7Mit66	57	62±19.9	53±22.2	52±20.4	0.06	0.72	1
D7Mit206	60	62±18.7	55±20.9	49±23.4	0.004	1.72	1
D7Mit105	63.5	64±22.8	53±17.2	48±24.0	0.001	2.16	0.5
D7Mit259	72	60±23.3	54±18.4	51±24.1	0.04	0.84	1
D8Mit58	1	44±17.1	57±23.8	59±15.7	0.002*	1.94	0.8
D8Mit94	13	45±16.8	57±23.5	58±16.7	0.03*	0.98	1
D8Mit63	15	45±16.8	57±23.7	57±16.3	0.04	0.9	1
D9Mit303	35	55±19.0	50±22.2	63±19.4	0.004*	1.71	1
D9Mit104	35	55±19.0	50±22.2	65±18.8	0.001*	2.09	0.6
D9Mit262	41	52±19.8	48±18.3	69±21.7	0.00002*	3.88	0.0139
D9Mit236	43	53±19.9	47±18.4	69±21.5	0.00002*	3.84	0.0153
D9Mit133	43	55±18.7	49±22.3	66±17.9	0.0003*	2.77	0.1

Table 4 (continued)

Marker[a]	Map location (cM)[b]	Colony number			P value[g]	LOD[h] score	Genome wide P value[i]
		aa[c]	*a/b*[d,e]	*bb*[f]			
D9Mit196	48	56±18.9	49±22.3	65±18.0	0.0006*	2.54	0.2
D9Mit273	49	56±18.8	49±22.2	65±18.2	0.0006*	2.52	1
D9Mit182	55	55±18.3	50±19.2	63±25.2	0.0256*	1.08	1

[a] Microsatellites markers.
[b] Map location (cM) from the centromere.
[c] *a/a*– homozygous genotype with two copies of the BALB/c allele.
[d] Mean number of colonies±SD.
[e] *a/b*– heterozygous genotype with one copy of the BALB/c allele and one copy of the C57BL/6 allele.
[f] *b/b*– homozygous genotype with two copies of the C57BL/6 alleles.
[g] Single-point significance levels. Number of colonies was compared between genotypes using the Wilcoxon rank sum test (two sides). *P* value was calculated using the program Qlink.
[h] LOD equivalent scores. The individual LOD equivalent score were calculated as described by Lander and Kruglyak (1995).
[i] Genomewide *P* value < 0.05 significant for whole genome.
* Single-point combined *P* value of colony number associated with one homozygous genotype (two C57BL/6 alleles or two BALB/c alleles) and corresponding heterozygous genotype (one C57BL/6 allele and one BALB/c allele) that have no significant differences (degrees of freedom=1).

Table 5 Association of small keratinocyte colony number in (C57BL/6×CB6F1) backcross mice

Marker[a]	Map location[b]	Colony number[c]		P value[f]	LOD score[g]
		bb[d]	*ab*[e]		
D8Mit58	1	44±28.2	29±13.7	0.04	0.84
D8Mit94	13	45±29.4	29±15.4	0.02	1.06
D8Mit339	23	50±31.87	28±14.3	0.02	1.10
D9Mit262	41	46±28.9	28±12.8	0.04	0.84
D9Mit133	43	48±29.3	25±11.2	0.004	1.78
D9Mit273	49	47±27.8	26±12.2	0.004	1.74
D9Mit182	55	45±28.4	28±13.3	0.03	0.98
D9Mit51	61	47±30.3	29±12.7	0.05	0.80

[a] Microsatellite markers.

[b] Map location (cM) from the centromere.

[c] Mean number of colonies±SD.

[d] *b/b* homozygous genotype with two copies of the C57BL/6 alleles.

[e] *a/b* heterozygous genotype with one copy of the BALB/c allele and one copy of the C57BL/6 allele.

[f] Single-point significance levels. Number of colonies was compared between genotypes using the Wilcoxon Rank Sum test (two sides). P value was calculated using the program Qlink.

[g] LOD equivalent scores. Individual LOD equivalent scores were calculated as described by Lander and Kruglyak (1995).

mid-portion of the chromosome contained the gene(s) involved in regulating the number of keratinocyte stem cells and the small KCs. Association of an increased number of small KCs with inheritance of the C57BL/6 allele was detected for the region between D9Mit262 and D9Mit236 loci with single-point P=0.00002 [log of the odds (LOD) score=3.88; genomewide P value=0.0139] and P=0.0002 (LOD score=3.84; P value for whole genome=0.0153) respectively (Table 4). A defined LOD score (Fig. 9, Tables 5 and 6) suggested a single locus.

Genotyping of 45 (C57BL/6×CB6F1) backcross mice that had been characterized for number of small KCs also confirmed linkage of the C57BL/6 alleles with number of small colonies between the D9Mit262, D9Mit133 and D9Mit273 loci, with single-point P=0.0481, 0.0041 and 0.0046, respectively (Table 5). Thus, two independent crosses indicated linkage of the C57BL/6 allele in the same region of mouse chromosome 9. Linkage was strongest (genomewide significance) between the central

Table 6 Loci associated with large keratinocyte colony number in intercross between BALB/c and C57BL/6 mice

Marker[a]	Map location (cM)[b]	Colony number[c]			*P* value[d]	LOD score[e]	Genome wide *P* value[f]
		aa[g]	*a/b*[h]	*bb*[i]			
D4Mit235	1.9	14±7.7	9±7.8	7±4.8	0.0002*	2.88	0.1
D4Mit264	1.9	14±7.7	9±7.9	7±4.8	0.0001*	3.05	0.08
D4Mit181	2.5	14±7.7	9±7.9	7±4.8	0.0001*	3.05	0.08
D4Mit2	6.5	14±7.	10±8.3	7±4.9	0.0004	2.4	0.3
D4Mit108	12.1	13±7.3	8±6.7	9±8.7	0.01	1.2	1
D4Mit151	28.6	11±7.4	9±7.2	8±8.3	0.03	0.99	1

[a] Microsatellites markers.
[b] Map location (cM) from the centromere.
[c] Mean number of colonies±SD.
[d] Single-point significant levels. Number of colonies was compared between genotypes using the Wilcoxon Rank Sum test (two sides). *P* value was calculated using the program Qlink.
[e] LOD equivalent scores. Individual LOD equivalent score were calculated as described by Lander and Kruglyak (1995).
[f] Genomewide *P* value <0.05 is significant for the whole genome.
[g] *a/a*– homozygous genotype with two copies of the BALB/c allele.
[h] *a/b*– heterozygous genotype with one copy of the BALB/c allele and one copy of the C57BL/6 allele.
[i] *b/b*–homozygous genotype with two copies of the C57BL/6 alleles.
* Single-point combined *P* value of colony number associated with two C57BL/6 alleles and corresponding heterozygous genotype (one C57BL/6 allele and one BALB/c allele) that have no significant differences (degrees of freedom=1).

and proximal region of chromosome 9. The locus D9Mit262 showed the highest level of significant association. Thus we conclude that the gene regulating the number of small colonies is situated in the D9Mit262 region and we designate this locus keratinocyte stem cell locus 1 (Kscl).

Suggestive evidence (Lander and Kruglyak 1995) for linkage with the number of small colonies was found with markers on mouse chromosomes 1, 6, 7, and 8 (Tables 4 and 5). On chromosome 1, the D1Mit178 locus had a LOD score of 1.65 and single-point *P*=0.0058. Analysis of 12 markers on this chromosome suggests that a gene influencing the number of small colonies may reside in a 35-cM interval from D1Mit178 to D1Mit132. A third wide region showing an association with number of small colonies with C57BL/6 alleles was found on Chromosome 6 between D6Mit93 and D6Mit323. A peak suggestive LOD score of 2.25 (single-point *P*=0.001) was obtained at D6Mit99. A fourth region with sug-

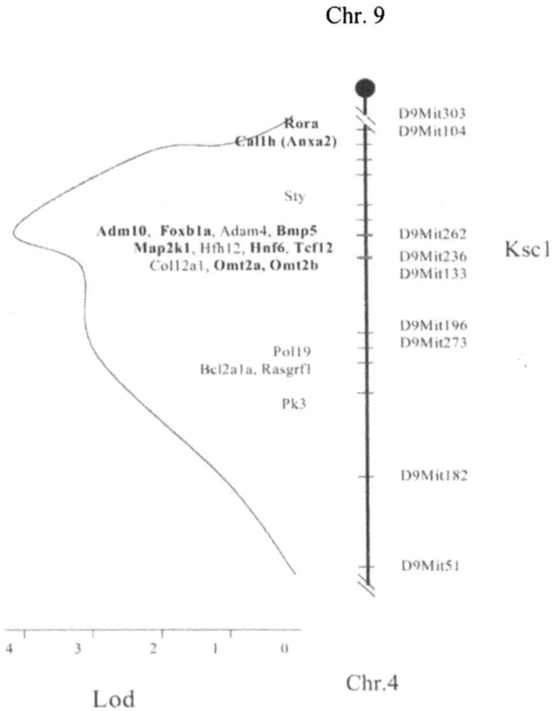
Chr. 9
Rora
Cal1h (Anxa2)
Sty
Adm10, Foxb1a, Adam4, Bmp5
Map2k1, Hfh12, Hnf6, Tcf12
Col12a1, Omt2a, Omt2b
Pol19
Bcl2a1a, Rasgrf1
Pk3
D9Mit303
D9Mit104
D9Mit262
D9Mit236
D9Mit133
D9Mit196
D9Mit273
D9Mit182
D9Mit51
Ksc1
4 3 2 1 0
Lod

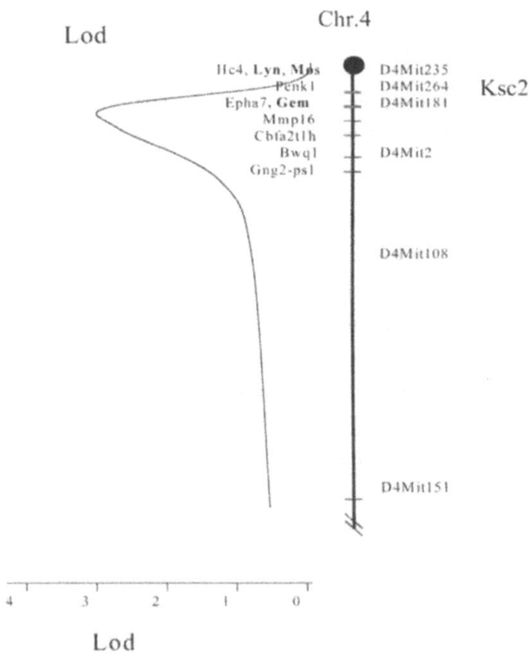
Chr.4
Hc4, Lyn, Mos
Penk1
Epha7, Gem
Mmp16
Cbfa2t1h
Bwq1
Gng2-ps1
D4Mit235
D4Mit264
D4Mit181
D4Mit2
D4Mit108
D4Mit151
Ksc2
4 3 2 1 0
Lod

gestive linkage to a high number of small colonies was identified between D7Mit66 and D7Mit259. A peak LOD score of 2.16 (single-point P=0.0016) was obtained at D7Mit105 (63.5 cM), and a peak LOD score of 1.94 (single-point P=0.002) was obtained at locus D8Mit58 near the centromere of chromosome 8. In the (C57BL/6×F1) backcross, the high number of small colonies was linked with C57BL/6 alleles in the same region of chromosome 8 between D8Mit58 and D8Mit339 (Table 5).

Suggestive linkage to large colony number was obtained for the markers on mouse chromosome 4 (Table 6). D4Mit264 and D4Mit181 had a LOD score of 3.05 (single-point P=0.0001). Analysis of eight markers on chromosome 4 suggested that the gene determining a high number of large colonies was situated between D4Mit264 (1.9 cM) and D4Mit181 (2.5 cM) and was associated with BALB/c alleles (Table 6). Thus a gene regulating the high number of large colonies resides in the D4Mit264 region. We designate this locus keratinocyte stem cell locus 2 (Ksc2).

3.4 Keratinocyte Stem Cells and Two-Stage Skin Carcinogenesis

BALB/c and C57BL/6 mice are almost equally resistant to two-stage skin carcinogenesis; however, several factors, including in vivo treatment with the carcinogen 7,12-dimethylbenz(a)anthracene (DMBA) and the promoter 12-*O*-tetradecanoylphorbol-13-acetate (TPA) involved in two-stage skin carcinogenesis, might also influence the number of clonogenic cells (Morris et al. 1988). Comparison of the loci associated with small and large KC number with published loci associated with resistance or susceptibility to two-stage skin carcinogenesis revealed overlap between

Fig. 9 Genetic linkage mapping of the number of small keratinocyte colonies with the Ksc1 locus on mouse chromosome 9, and the number of large keratinocyte colonies with the Ksc2 locus on mouse chromosome 4, based on analysis of 104 (BALB/c×C57BL/6) F2 female mice at 7 weeks of age. Microsatellite markers are shown on the *right* of the chromosome. LOD score plots for the number of small and large keratinocyte colonies are shown at the *bottom*. Two small regions on chromosomes 9 and 4 (*bold lines*) demonstrated the highest associations with small or large colonies. Genes mapping to these two regions are shown to the *left* of the chromosomes. Candidate genes for involvement in the regulation of keratinocyte stem cell number are in the *boldface*

a locus involved in skin tumor promotion, Psl1, previously mapped to a region of chromosome 9 (Angel et al. 1997), and the Ksc1 locus. In particular, inheritance of the DBA/2 allele on chromosome 9 is associated with increased susceptibility to TPA promotion (Angel et al. 1997). C57BL/6 mice represent the extreme of resistance to skin TPA tumor promotion. Unfortunately, Angel et al. (1997) reported the combined data for Psl1 in backcross, intercross and recombinant strains, making it difficult for us to conduct a more detailed analysis of Psl1 in association with Ksc1. Moreover, Mock et al. (1998) reported a higher than expected linkage score for the marker D9Mit271 (48 cM) in association with skin tumor susceptibility. In that study, inheritance of the BALB/c allele on the central region of mouse chromosome 9 represented resistance, while inheritance of SENCARA/Pt alleles of that region was associated with sensitivity to papilloma development. This locus is also closely related to our Ksc1. Thus a gene on chromosome 9 regulating the number of small KCs may also affect sensitivity/resistance to skin tumor promotion.

Nagase et al. (1995) identified two loci (Skts1, Skts2) on chromosome 7 that are linked to papilloma development. The Mus spretus allele of Skts2 (64 cM) in the distal region of chromosome 7 is associated with female-specific papilloma resistance, while the *Mus musculus* allele in this locus is associated with susceptibility to papilloma development. In our intercross mice, the BALB/c allele of D7Mit105 (63.5 cM) locus showed suggestive association with the high number of small colonies in almost the same region as Skts2. Thus, a gene regulating small KC number might be involved in the regulation of susceptibility or resistance to papilloma development in the Skts2 locus.

A large region on chromosome 6 linked with susceptibility or resistance to benign and malignant skin tumors (Nagase et al. 1995) correlated with the region on chromosome 6 that was associated with a high number of small colonies (Table 4). Our LOD score peak on chromosome 6 (D6Mit99, 36 cM) (Table 4) was also situated in the region close to the peak *P* value on chromosome 6 (D6Mit9, 32 cM; D6Mit29, 35 cM) mapped by Nagase et al. (1995). The gene for transforming growth factor-α situated on chromosome 6 at 35.8 cM from the centromere has been implicated in papilloma development (Vassar et al. 1992) and might also influence keratinocyte colony number (R.J. Morris, unpublished results). Together, the data suggest that a gene(s) linked with the number of small colonies on chromosome 6, also influences the development of benign and malignant skin tumors. Finally, the Kscl locus on chromosome 9 and two minor suggestive loci on chromosomes 7 and 6 are surprisingly close to loci found to be associated with skin tumor promotion (Angel et al. 1997)/sensitivity/resistance to skin tumor development (Nagase et al. 1995; Mock et al. 1998). It is possible that the population of clonogenic keratinocytes able to produce small colonies represents a specific population of skin stem cells responsible for sensitivity or resistance to skin carcinogenesis. Based on our data, skin stem cells resistant to carcinogenesis appear to be associated with genes in the C57BL/6 alleles on chromosomes 9 and 6, and the BALB/s allele on chromosome 7. Identification of loci responsible for sensitivity to skin carcinogenesis awaits further studies of other genetic crosses with strains susceptible to skin carcinogenesis.

The Kscl region of mouse chromosome 9 (Fig. 9) shares linkage homology with human chromosomes 15q22.1 (MAP2K1), 15q21 (TCF12), 15q21-q22 (ADAM10), 6pter-qter (BMP5), 14pter-qter (COX7A3). The region of the Kscl locus contains a cluster of genes responsible for normal mouse development (Adam10, 4; Foxb1a; Bmp5; Hnf6; Tcf12; Om2a, 2b) and possibly involved in regulation of keratinocyte stem cell number through their effects on growth and differentiation (Neurman et al. 1993; Sasaki and Hogan 1993; West et al. 1996; Wolfsberg and White 1996; Furuta et al. 1997; Yavari et al. 1998; Jacquemin et al. 2000). Interestingly, Adam10 plays a role in the Notch signaling pathway, which has been implicated in keratinocyte growth arrest and entry into differentiation (Rangarajan et al. 2001), and may be expressed in the mouse hair follicle (Kopan and Weintraub 1993). The involvement of Delta-Notch signaling in stimulating human epidermal stem cell differentiation (Lowell et al. 2000), further suggests Adam 10 as a candidate gene in the Kscl locus.

Another group of genes in the Kscl region includes those encoding mitogen-activated protein kinase 1 (Map2k1) (Alessandrini et al. 1997) and Anxa2, a major substrate for growth factor receptor protein-tyrosine kinases and protein kinase C (Saris et al. 1987). Morris et al. (1988) showed that the number of clonogenic keratinocytes increases in response to TPA treatment, possibly resulting in protein kinase signal transduction. Thus, Map2k1or/and Anaxa2 might be involved in the regulation of the number of clonogenic keratinocytes.

The Ksc2 locus on mouse chromosome 4 shares linkage homology with human chromosomes 8q13 (LYN, GWM), 8q11 (MOS), 8q13-q21 (GEM). Possible candidate genes for Ksc2 include Lyn, a member of the

Src subfamily encoding protein tyrosine kinase and expressed in skin (Hibbs et al. 1995), and Mos, a post-transcriptionally regulated gene expressed during oocyte growth and maturation (Newman and Dai 1996). Finally the mitogen-induced gene, Gem, encodes a GTP-binding protein that belongs to a new family within the Ras superfamily (Santoro et al. 1995); the regulated expression pattern of Gem suggests a role for this protein in cellular responses to growth stimulation and it may be a candidate gene for the Ksc2 locus.

4 Conclusion

Our data demonstrate that keratinocyte stem cell number is a genetically defined and quantitative complex trait. Two major subpopulations of clonogenic keratinocyte stem cells are regulated by different genes, with the locus responsible for small colony number on mouse chromosome 9 and designated keratinocyte stem cell locus 1 (Kscl), and the locus responsible for large colonies on chromosome 4 and designated keratinocyte stem cell locus 2 (Ksc2). Correlative data suggest that clonogenic keratinocytes producing small colonies might represent a specific population of keratinocytes stem cells responsible for sensitivity or resistance to skin carcinogenesis.

The identification of genes regulating amplification of keratinocyte stem cells awaits the identification and eventual cloning of the Kscl and Ksc2 genes. Moreover, the mechanisms regulating the intrinsic number of keratinocyte stem cells, possibly part of a universal regulatory mechanism for maintaining resting numbers of stem cells or their amplification, undoubtedly underlie the cells responses to external manipulation. The ability to manipulate these genes in vivo raises the exciting possibility of a novel treatment for skin disease and cancer. Further studies are needed to determine whether the stem cell regulatory genes might be targets in carcinogenesis caused by environmental chemicals or ultraviolet light.

Acknowledgments. This work was supported in part by NIH grant CA45293 (R.J.M.) and by the Lankenau Foundation. We thank Dr. Spencer M. Free for statistical analysis.

References

Alessandrini A, Brott BK, Erikson RL (1997) Differential expression of MEK1 and MEK2 during mouse development. Growth Differ 8:505–511

Angel J, Popova N, Lanko N, Turusov V, DiGiovanni J. (2000) A locus that influences susceptibility to 1,2-Dimethylhydrazine-induced colon tumors maps to the distal end of mouse Chromosome 3. Mol Carcinogen 27:47–54

Angel J, Beltran L, Minda K, Rupp T, DiGiovanni J (1997) Association of murine chromosome 9 locus (Psl1) with susceptibility to mouse skin tumor promotion by 12-O-Tetradecanoylphorbol-13-acetate. Mol Carcinogen 20:162–167

Beck J A, Lloyd S, Hefezparasr M, Lennon-Pierce M, Eppig JT, Festing MFW, Fisher EMC (2000) Genealogies of mouse inbred strains. Nat Genet 24:23–25

Box GEP, Hunter WG, Hunter JS. (1978) Statistics for experimenters. In: An Introduction to Design, Data Analysis, and Model Building. John Wiley & Sons. New York, NY

Cotsarelis G, Kaur P, Dhouailly D, Hengge U, Bickenbach J (1999) Epidermal stem cells in the skin: definition, markers, localization and functions. Exp Dermatol 8:80–8

Elston RC, Stewart J (1973) The biomedical genetics with one or two loci: the inheritance of physiological characters in mice. Genetics 73:675–693

Green EL (1981) Segregation of alleles genetics and probability. In: Genetics and Probability in Animal Breeding Experiments. Macmillan Publishers Ltd, London, pp 50–59

Furuta Y, Piston D, Hogan B (1997) Bone morphogenetic proteins (BMPs) as regulators of dorsal forebrain development. Development 124:2203–2212

Fuchs E, Raghavan S (2002) Getting under the skin of epidermal morphogenesis. Nature Rev Genet 3:199–209

Jacquemin P, Durviaux SM, Jensen J, Godfraind G, Guillemot F, Madsen OD, Carmeliet P, Dewerchin M, Collen D, Rousseau GG, Lemaigre FP (2000) Transcription factor hepatocyte nuclear factor 6 regulates pancreatic cell differentiation and controls expression of the proendocrine gene ngn3. Mol Cell Biol 20:4445–4454

Hibbs M, Stanley E, Maglitto R, Dunn A (1995) Identification of a duplication of the mouse gene. Gene 24: 175–181

Kopan R, Weintraub H (1993) Mouse Notch: expression in hair follicles correlated with cell fate determination. J Cell Biol 121:631–641

Lander E, Kruglyak L (1995) Genetic dissection of complex traits: guidelines for interpreting and reporting linkage results. Nat Genet 11:241–247

Lowell S, Jones P, Le Roux I, Dunne J, Watt FM (2000) Stimulation of human epidermal differentiation by Delta-Notch signaling at the boundaries of stem-cells clusters. Curr Biol 10:491–500

McBrearty BA, Clark LD, Zhang XM, Blankenhorn EP, Heber-Katz E (1998) Genetic analysis of a mammalian wound-healing trait. Proc Natl Acad Sci USA 95:11792–11797

Mock B, Lowry D, Rehman I., Padlan C, Yuspa S, Hennings H (1998) Multigenic control of skin tumor susceptibility in SENKAR/Pt mice. Carcinogenesis 19:1109–1115

Kobayashi K, Rochat A, Barrandon Y (1993) Proc Natl Acad Sci USA 90:7391–7395
Morris RJ (1994) Procedure for harvesting epidermal cells from the dorsal epidermis of adult mice for primary cell culture in 'high calcium' defined medium. In: Leigh I. and Watt F (eds) Keratinocyte Methods, Cambridge University Press, pp 25–31
Morris RJ, Potten CS (1994) Slowly cycling label-retaining epidermal cells behave like clonogenic stem cells in vitro. Cell Prolif 27:279–289
Morris R, Potten C (1999) Highly persistent label-retaining cells in the hair follicles of mice and their fate following induction of anagen. J Invest Dermatol 112:470–475
Morris RJ, Tacker KC, Fischer SM, Slaga TJ (1988) Quantitation of primary in vitro clonogenic keratinocytes from normal adult murine epidermis following initiation and during promotion of epidermal tumors. Cancer Res 48:6285–6290
Morris RJ, Fisher SM, Slaga TJ (1985) Evidence that the central and peripherally located cells in the murine epidermal proliferative unit are two distinct cell populations. J Invest Dermatol 84:277–281
Nagase H, Bryson S, Cordell H, Kemp C, Fee F, Balmain A (1995) Distinct genetic loci control development of benign and malignant skin tumors in mice. Nat Genet 10:424–429
Newman B, Dai Y (1996) Transcription of c-mos protooncogene in the pig involves both tissue-specific promoters and alternative polyadenylation sites. Mol Reprod Dev 44:275–288
Neurman T, Keen A, Knapik E, Shain D, Ross M, Nornes HO, Zuber MX (1993) ME1 and GE1: basic helix-loop-helix transcriptional factors expressed at high levels in the developing nervous system and in morphogenetically active regions. Eur J Neurosci 5: 311–318
Oshima H, Rochat A, Kedzia C, Kobayashi K, Barrandon Y (2001) Morphogenesis and renewal of hair follicles from adult multipotent stem cells. Cell 104:233–245
Popova N, Tryson K, Wu K, Morris R (2002) Evidence that keratinocyte colony number is genetically controlled. Exp Dermatol 11:503–508
Popova NV, Teti KA, Wu K, Morris RJ (2003) Identification of Two Keratinocyte Stem Cell Regulatory LociImplicated in Skin Carcinogenesis Carcinogenesis 24:417–425
Potten CS (1986) Cell cycles in cell hierarchies. Int J Radiat Biol Relat Stud Phys Chem Med 49:257–78
Rangarajan A, Talora C, Okuyama R, Nicolas M, Mammucari C, Oh H, Aster J, Krishna S, Metzger D, Chambon P, Miele L, Aguet M, Radtke A, Dotto, GP (2001) Notch signaling is a direct determinant of keratinocyte growth arrest and entry into differentiation. EMBO J 20: 3427–3436
Rochat A, Kobayashi K, Barrandon Y (1994) Location of stem cells of human hair follicles by clonal analysis. Cell 76:1063–1073
Santoro T, Maguire J, McBrige O, Avraham K, Copeland N, Jenkins N, Kelly K (1995) Chromosomal organization and transcriptional regulation of human GEM and localization of the human and mouse GEM loci encoding an inducible Ras-like protein. Genomics 10:558–564
Saris CJ, Kristensen T, D'Eustachio P, Hicks LJ, Noonan DJ, Hunter T, Tack BF (1987) cDNA sequence and tissue distribution of the m-RNA for bovine and murine p11,

the S100-related light chain of the protein-tyrosine kinase substrate p36 (capactin 1). J Biol Chem 5:10663–10671

Sasaki H, Hogan BL (1993) Differential expression of multiple fork head related genes during gastration and axial pattern formation in the mouse embryo. Development 118:47–59

Vassar R, Hutton ME, Fuchs E (1992) Transgenic overexpression of transforming growth factor alpha bypasses the need for c-Ha-ras mutation in mouse skin tumorigenesis. Mol Cell Biol. 12:4643–4653

Watt FM (2001) Stem cell fate and patterning in mammalian epidermis. Curr Opin Genet Dev 11:410–7

West MF, Verrotti AC, Salles FJ, Tsirka SE, Strickland S (1996) Isolation and characterization of two novel, cytoplasmically polyadenylated, oocyte-specific, mouse maternal RNAs. Dev Biol 175:132–141

Wolff R, Gemmill R (1997) Purifying and analyzing genomic DNA. In: Birren B, Green E, Klapholz S, Myers R, Roskams J (eds). Genome analysis. A laboratory manual. Vol 1. New York: Cold Spring Harbor Laboratory Press, pp 1–82

Wolfsberg TG, White JM (1996) ADAMs in fertilization and development. Dev Biol 15:389–401

Yavari R, Adida C, Bray-Ward P, Brines M, Xu T, (1998) Human metalloprotease-disintegrin Kuzbanian regulates sympathoadrenal cell fate in development and neoplasia. Hum Mol Genet 7:1161–1167

Zhu AJ, Hoose I, Watt F (1999) Signalling via $\beta 1$ integrins and mitogen-activated protein kinase determines human epidermal stem cell fate in vitro. Pros Natl Acad Sci USA 96:6728–6733

Myocardial Tissue Engineering and Regeneration as a Therapeutic Alternative to Transplantation

A. S. Krupnick · D. Kreisel · M. Riha · K. R. Balsara · B. R. Rosengard
Department of Surgery, Division of Cardiothoracic Surgery, Hospital of the University of Pennsylvania, 6 Silverstein Building, 3400 Spruce Street, Philadelphia, PA 19104, USA
E-mail: krupnica@uphs.upenn.edu

1 Introduction 140

2 Justification for Research Focusing on Myocardial Tissue Engineering . . 140

3 Cellular Cardiomyoplasty 142
3.1 Cellular Cardiomyoplasty Utilizing Cardiac Myocytes 142
3.2 Cellular Cardiomyoplasty Utilizing Skeletal Myoblasts 143
3.3 Cellular Cardiomyoplasty Utilizing Stem/Progenitor Cells 143
3.4 Clinical Applications and Physiologic Relevance of Cellular Cardiomyoplasty 144

4 Pediatric Applications 145
4.1 Small Animal Model of Ventricular Tissue Engineering 147
4.2 Autologous Source of Progenitor Cells for Perinatal and Fetal Cardiac Tissue Engineering 149

5 Current Limitations and Future Directions of Cardiac Tissue Engineering 152

6 Myocardial Regeneration 153

7 Conclusions 158

References 159

Abstract Ischemic cardiomyopathy leading to congestive heart failure remains the leading source of morbidity and mortality in Western society and medical management of this condition offers only palliative treatment. While allogeneic heart transplantation can both extend and improve the quality of life for patients with end-stage heart failure, this therapeutic option is limited by donor organ shortage. Even after successful transplantation, chronic cardiac rejection in the form of cardiac allograft vasculopathy can severely limit the lifespan of the transplanted organ. Current experimental efforts focus on cellular cardiomyoplasty, myocardial tissue engineering, and myocardial regeneration as alternative approaches to whole organ transplantation. Such strategies may offer novel forms of therapy to patients with end-stage heart failure within the near future.

1 Introduction

Over the last several decades tremendous advances have been made in the field of allogeneic cardiac transplantation and immunosuppression but donor organ shortage remains a major limitation to this form of therapy. Aside from the lack of suitable organs for transplantation our recent description of a powerful alternative pathway of allorecognition, mediated solely by graft parenchymal cells, established the existence of yet another barrier to successful long-term allogeneic cardiac transplantation. These issues have forced us to reexamine the necessity of whole organ transplantation and consider the possibility that replacement of only the portion of myocardium lost to ischemic insult or congenital malformation might suffice in treating cardiac dysfunction. Despite long-standing assumptions that adult myocardium cannot regenerate and lacks progenitor or stem cells to support such regeneration, recent experimental data has shown that adult and fetal stem cells can differentiate into cardiac myocytes, repopulate myocardial scar, and improve myocardial function. Based on these findings tremendous experimental efforts have focused on the establishment of small animal models of adult and congenital cardiac disorders in order to evaluate novel cell delivery strategies such as cellular cardiomyoplasty, myocardial tissue engineering, as well as to harness innate regenerative capacity of myocardial tissue. Such transitional research offers the opportunity to translate basic stem cell biology into therapeutic clinical applications within the near future.

2 Justification for Research Focusing on Myocardial Tissue Engineering

Despite tremendous advances in the treatment of acute myocardial infarction, post-infarction congestive heart failure remains a difficult clinical problem (Tsuyuki 2001). Coronary occlusion with consequent regional ischemia leads to loss of cardiomyocytes and progressive replacement of muscle by collagenous tissue resulting in a myocardial scar. Even a small scar progressively expands, directly affecting the contractile function of adjacent border zone myocardium and overall global contractile function. Thus, left ventricular dilatation and heart failure can occur even after a myocardial infarction of moderate size (Pfeffer 1991; Kelley 1999). While novel therapeutic strategies such as biventricular pacing, mitral valve repair and ventricular remodeling surgery may improve the quality of life in a portion of patients with end-stage heart failure, heart transplantation is still the only available treatment that significantly lengthens life expectancy. Transplantation, however, is limited by a chronic shortage of donor hearts.

Advances in immunosuppression have significantly improved the treatment of acute rejection but chronic rejection in the form of cardiac allograft vasculopathy remains a leading cause of death in transplant recipients (Hollenberg 2001). Organ rejection can be initiated by two separate pathways of allorecognition. Unlike physiologic 'indirect' T-cell mediated allorecognition, where a foreign antigen (in this case an alloantigen) is processed and presented by a recipient antigen presenting cell to a recipient T cell in an MHC self-restricted manner, 'direct allorecognition' is driven by the recognition of the intact alloantigen on an allogeneic antigen presenting cell. Due to this molecular mimicry, the precursor frequency of alloreactive T cells activated by this pathway is several logs higher than those that can be activated by indirect allorecognition alone (Benichou et al. 1999; Suchin et al. 2001). The vigor of the initial episode of acute rejection after organ transplantation has been classically attributed to donor-derived hematopoietic antigen presenting cells, which migrate from the transplanted organ to the host's peripheral lymphoid tissue and activate T lymphocytes by the direct pathway of allorecognition. As these passenger leukocytes survive only briefly after migration to the host lymphoid system, allorecognition at later points and subsequential chronic rejection has been attributed to the weaker 'indirect pathway' (Ensminger et al. 2002; Lee et al. 2001).

To evaluate some inconsistencies in this theory our laboratory focused on the role of allograft parenchymal cells in direct allorecognition. We found that even in the absence of professional donor-derived hematopoietic antigen presenting cells donor parenchymal cells, most likely vascular endothelial cells, can initiate direct alloantigen presentation to alloreactive host T lymphocytes. In fact the presence of donor derived hematopoietic antigen presenting cells does not affect the tempo of $CD8^+$ T-cell mediated allograft rejection which can be mediated entirely by parenchymal cells (Kreisel et al. 2002). Thus, a cell population that resides permanently in the transplanted organ and plays a critical role in organ function also can initiate a powerful pathway of antigen presentation leading to chronic rejection and organ destruction. While these findings will eventually lead to an improvement in allograft survival by allowing the development of immunosuppression and costimulatory blockade agents targeting this pathway, this new paradigm of allorecognition presents yet another major barrier to successful, long-term, solid organ transplantation. But is whole organ replacement really necessary for treatment of end-stage heart failure? Since the majority of heart failure patients suffer a decline in function due to deleterious ventricular remodeling resulting from a limited size infarction, it may be possible to prevent heart failure by restoring contractility solely in this portion of the myocardium.

3
Cellular Cardiomyoplasty

Microvascular repair of infarcted myocardium is characterized by neoangiogenesis, with eventual ingrowth of vasculature into myocardial scar, while damaged ventricular myocytes are mostly replaced by noncontractile cells. The technique of cellular cardiomyoplasty involves direct localized injection of contractile cells into myocardial scar in order to convert the dyskinetic scar into functional tissue.

3.1
Cellular Cardiomyoplasty Utilizing Cardiac Myocytes

Initial experimental attempts utilized syngeneic fetal cardiac myocytes, which were transplanted into uninjured, adult myocardium (Soonpaa et al. 1994). This study provided encouraging results with stable long-term engraftment of the transplanted cells. Following on this effort numerous investigators applied such techniques to scar tissue. The results were overwhelmingly positive and the majority of the investigators report an improvement in overall ventricular function, reversal of heart failure, and improvement in survival (Leor et al. 1996; Li et al. 1996a, 1996b, 1997; 2000; Scorsin et al. 1996, 1997; Reinecke et al. 1999; Sakai et al.

1999; Yoo et al. 2000; Yokomuro et al. 2001). The clinical applications of cardiac myocyte therapy, however, are limited in the human by donor site morbidity associated with harvest of autologous cardiac myocytes, the difficulty of in vitro expansion of such cells and requirement for immunosuppression if an allogeneic cell source is utilized.

3.2 Cellular Cardiomyoplasty Utilizing Skeletal Myoblasts

Unlike cardiac muscle, the harvest of peripheral skeletal muscle is associated with minimal patient morbidity. Furthermore, skeletal muscle, even in the adult, contains undifferentiated myoblasts and satellite cells intimately associated with the myofibers. Based on these principles several groups evaluated the feasibility of utilizing this cell population for post-infarction cellular cardiomyoplasty (Taylor 1998; Pouzet 2000). Similar to cardiac myocytes, skeletal myoblasts survived, incorporated into the myocardial scar and established intracellular connections through the formation of intercalated disks. More importantly some cells developed phenotypic characteristics of cardiac myocytes with associated improvement in overall global ventricular function (Taylor 1998).

3.3 Cellular Cardiomyoplasty Utilizing Stem/Progenitor Cells

The ideal candidate for cellular cardiomyoplasty, however, is likely to be a less committed cell, which can undergo full cardiogenic differentiation. While traditional dogma has held that the adult heart lacks a reserve of stem cells, recent developments have placed these assumptions in question. Mesenchymal progenitor or stem cells have been isolated from the adherent bone marrow stroma of numerous species, including man, and have shown a strong propensity for in vitro expansion and differentiation (Pittenger et al. 1999). Systemic transplantation of this cell population, even across xenogeneic barriers, has been used to document the cardiomyocytic differentiation potential of these cells (Liechty et al. 2000; Toma et al. 2002). Similar to the results obtained with skeletal myoblasts, transplantation of this cell population into myocardial scar resulted in site-specific differentiation and improvement in overall myocardial function (Tomita et al. 1999). Interestingly, cardiomyocytic differentiation is not restricted to the stromal component of the bone marrow. Lin^{-}c-kit^{pos} bone marrow cells, traditionally thought to be extremely primitive hematopoietic stem cells, can also differentiate into cardiac myocytes and lead to an improvement in ventricular function (Orlic et al. 2001a). Such 'transdifferentiation' has been reported for even terminally differentiated cells such as vascular endothelium (Condorelli et al. 2001). This previously inconceivable plasticity of adult cells has stimulated reevaluation of therapeutic strategies for end-stage heart failure and greatly increased the number of laboratories studying cell therapy for the failing heart.

3.4 Clinical Applications and Physiologic Relevance of Cellular Cardiomyoplasty

Despite these initial successful results, only limited studies document systolic thickening and contractility within the area of cell transplantation (Atkins et al. 1999; Rajnoch et al. 2001) and most investigators document improvement in global cardiac function only. Introduction of viable cells into myocardial scar changes the material properties of the tissue, increases neoangiogenesis due to secretion of growth factors by the cellular inoculum, and changes the overall compliance of the scar. Therefore, cellular cardiomyoplasty might result in improvement of ventricular function on the basis of preventing scar expansion and aberrant ventricular remodeling, similar to that achieved by simple physical infarct restraint (Kelley 1999). The fact that transplantation of noncontractile cells such as fibroblasts into the area of the infarct also improves ventricular function, especially diastolic performance, lends further credence to such a notion (Li 1999; Sakai 1999; Hutcheson et al. 2000).

One must also consider the limitation of the delivery system that utilizes injection of cells directly into myocardial scar. As cells are distributed randomly within the site of the needle tract and 'trapped' within the scar matrix, it is unlikely that a large portion of them will properly migrate and organize into coordinated contractile tissue. Similar setbacks have been encountered in other therapeutic applications of myogenic cell therapy such as treatment of muscular dystrophy (Tremblay and Guerette 1997). Myocardial scar, which consists of a collagenous fibrous matrix, also lacks the three-dimensional extracellular structure necessary to tether myogenic cells together in order to coordinate synchronized contraction.

Despite these theoretical and experimental limitations, overall results of animal studies have been encouraging with improvement in ventricular function and animal survival reported in most models. The clinical demand combined with the current limitations of therapeutic treatments of end-stage heart failure has paved the way for the first human trial of autologous skeletal myoblast transplantation which was initiated in France during the year 2000. Although preliminary published data suggest improvement in ventricular performance and new onset metabolic activity in the area of myoblast transplantation, these conclusions need careful and systematic analysis, especially since cell therapy was combined with myocardial revascularization (Menasche et al. 2001). While no arrhythmogenic complications were reported in the first patient, caution must be exercised as the majority of the data studying this therapy have been acquired in rodent models, which are generally resistant to arrhythmias. The use of culture expanded cells must also be carefully evaluated as such treatment can alter the genetic stability with potential for malignant transformation and aberrant gene expression (Morshead et al. 2002). Lessens learned from the untimely death of a young man undergoing a human gene therapy trial at the University of Pennsylvania, and the resulting set back to the field of gene therapy due to his death, warrant careful optimism as well as controlled and vigorous animal experimentation prior to large scale human trials (Raper et al. 2002).

4 Pediatric Applications

Despite advances in the field of cellular cardiomyoplasty, the therapeutic applications of this technology are limited in the field of pediatric and neonatal heart surgery where congenital cardiac anomalies rather than ischemic cardiomyopathy are the leading cause of death (Srivastava and Olson 2000). Right or left ventricular hypoplasia with atresia of the aortic arch or pulmonary vasculature and associated valvular structures represent the most complex of such disorders (Fig. 1). While circulation with such univentricular physiology can proceed through the patent ductus arteriosus in utero, postnatal survival depends on palliative, staged reconstruction or transplantation. Transplantation in children is limited as a permanent solution due to the scarcity of size-matched organs and the problem of chronic rejection during the expected longevity of the child. Long-term results of palliative reconstruction, which per-

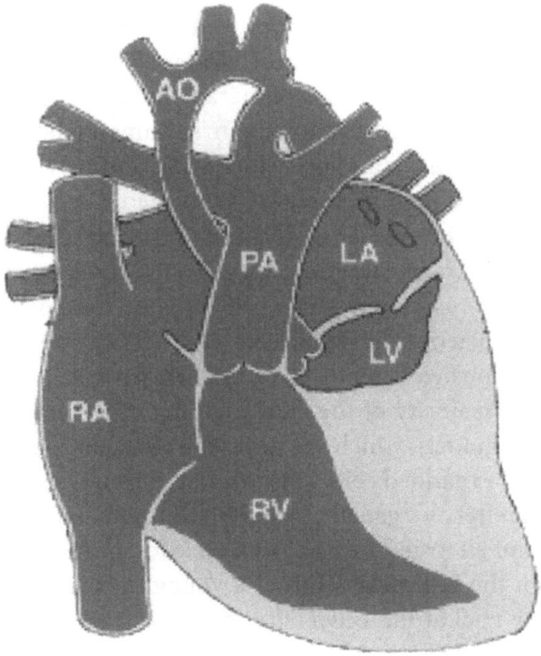

Fig. 1 Representative picture of hypoplastic left heart syndrome with atresia of the aortic arch, aortic valve, and left ventricular cavity. *AO*, aortic arch; *LV*, left ventricle; *LA*, left atrium; *RA*, right atrium; *RV*, right ventricle; *PA*, pulmonary artery

manently convert the circulation to univentricular physiology, are still unknown.

Tissue engineering is a multidisciplinary field combining principles of engineering and biological science and focuses on creating viable organs and tissues to address the limitations of allogeneic solid organ and cell transplantation (Langer and Vacanti 1993; Vacanti and Langer 1999). The basic principle involves in vivo implantation of a cell-matrix construct in order to replace diseased or deficient tissues with recapitulation of structure after resorption or remodeling of the matrix. Recent developments in successful tissue engineering of cardiac valves and blood vessels have opened up the possibility of reconstructing the hypoplastic left heart by replacing the atretic aortic arch and valve with autologous living constructs that can properly grow during the child's development (Shinoka et al. 1995; Niklason et al. 1999; Hoerstrup et al. 2000). Complete correction of cardiac anomalies, however, is limited by the inability to engineer functional ventricular tissue and enlarge the ventricular cavity. Manipulation of the native heart for tissue engineering purposes is possible in a large animal model under complete cardiopulmonary bypass, but no comparable technology is available for rodent models (Shinoka et al. 1996). This is unfortunate as the murine model offers the ideal opportunity to study in vivo myocardial tissue engineering with a substantial amount of in vitro data describing the construction of myocardial tissue already available (Carrier et al. 1999; Zimmermann et al. 2000). A rodent model also offers numerous other advantages such as the availability of inbred syngeneic strains, a defined system of stem cells for transplantation, as well as decreased animal cost.

4.1 Small Animal Model of Ventricular Tissue Engineering

To address this need we set out to develop a small animal model for the study of myocardial tissue engineering utilizing heterotopic heart transplantation. This model takes advantage of the fact that an explanted heart, after cold cardioplegia, can tolerate several hours of ischemia without a decrease in performance and if transplanted into a syngeneic animal can function for the life of that animal without immunologic rejection (Snabaitis et al. 1997). After a modification of microsurgical techniques we were able to vary the functional capacity of the left ventricle by manipulating the microsurgical anastomoses of the transplanted heart. In hearts transplanted in the 'standard' fashion with a donor to recipient aorto-aortic anastomosis and the donor pulmonary artery to recipient inferior vena caval anastomosis, the left ventricle is functionally excluded from the circulation and models the clinical scenario of left ventricular support provided by a mechanical assist device (Fig. 2A) (Ono and Lindsey 1969). When transplanted in a more physiologic fashion, by creating in-flow through a donor superior vena cava to recipient inferior vena cava anastomosis, and loading the left ventricle via a donor pulmonary artery to left atrial anastomosis, the left ventricle is forced to perform mechanical work (Fig. 2B). Unlike the native rat heart, which can generate an ejection fraction close to 65%, ejection fractions of only 20%–25% are recorded in this heterotopic heart as only a small portion of the blood flow is directed into the graft (Liu et al. 1997). Despite this limitation, manipulation, infarction and ventricular engineering can be

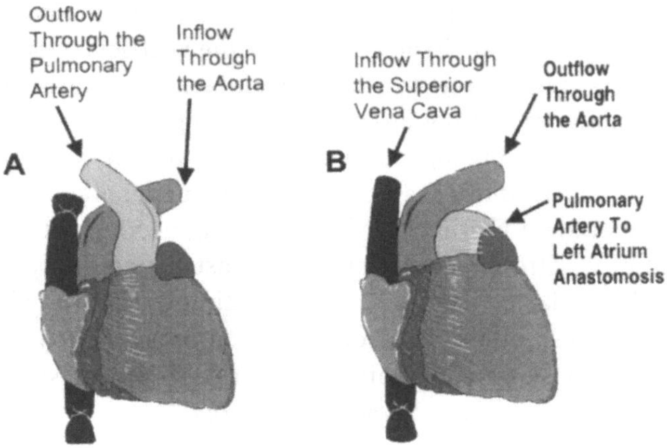

Fig. 2A, B Modification of microsurgical anastomoses can vary the functional loading of the left ventricle. **A** By anastomosing the donor ascending aorta to the recipient intra-abdominal aorta and the donor pulmonary artery to the recipient inferior vena cava, the left ventricle is excluded from the circulation. Oxygenated blood from the recipient enters the donor aorta, perfuses the myocardium through the coronary arteries but does not enter the left ventricle due to the competent aortic valve. This blood is then directed into the right ventricle through the coronary sinus and is ejected into the recipient inferior vena cava via the pulmonary artery. **B** By anastomosing the donor superior vena cava to the recipient inferior vena cava and creating a pulmonary artery to left atrium anastomosis a functioning left ventricle is created. Blood enters the heterotopic heart through both the superior vena cava and the aorta and is ejected via the aorta during systole. (Modified from Krupnick et al. 2002a)

performed on such heterotopic hearts with minimal animal loss. Novel experimental opportunities, which cannot be routinely performed on the rodent orthotopic heart, are feasible in this model (Krupnick et al. 2002a).

After establishing the microsurgical techniques, we set out to validate such an experimental approach as a model for neonatal ventricular tissue engineering. We studied various biocompatible constructs and found that a Matrigel/Type I collagen hydrogel supported by a porous polylactide biodegradable mesh and reinforced with polytetrafluoroethylene (Gore-Tex, WL Gore and Associates, Inc., Flagstaff, AZ) covering provided the optimal material for ventricular reconstruction. Unlike collagen hydrogels constructed without a porous scaffolding, the composite construct could be fashioned into various predetermined shapes and provided optimal handling characteristics necessary for suture place-

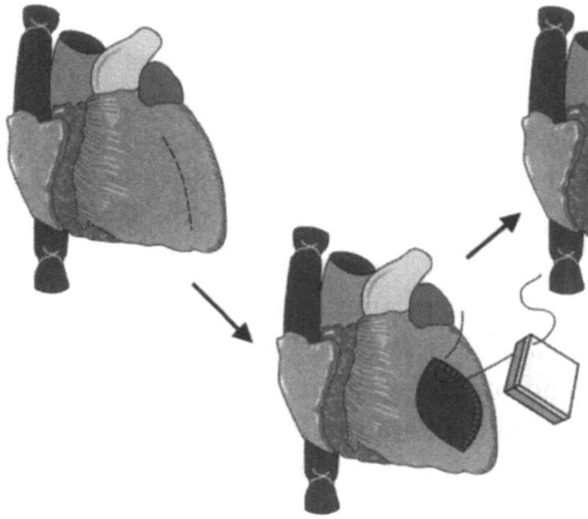

Fig. 3 The size of the ventricular cavity could be increased by implantation of biocompatible scaffolding. This scaffolding could also be utilized as a vehicle for delivery of cells to the ventricle. (Reproduced from Krupnick et al. 2002a)

ment into ventricular myocardium. Unlike several biodegradable materials evaluated, Gore-Tex reinforcement prevented distortion of ventricular geometry and aneurysmal dilatation of the functioning left ventricle. When 5-bromo-2-deoxy-uridine (BrdU) labeled multipotent adult bone marrow-derived mesenchymal progenitor cells, isolated from syngeneic animals, were seeded into the collagen hydrogel prior to transplantation, these cells survived and expressed cardiac specific cytoplasmic contractile proteins suggestive of cardiomyocytic differentiation. More importantly the size of the ventricular cavity, as measured by echocardiography, could be augmented by matrix implantation (Fig. 3). Thus, we had established a working rodent model for studying myocardial tissue engineering with an emphasis on congenital anomalies. Despite the theoretic potential of this model we still lacked an autologous and clinically relevant source of cells necessary to reconstruct myocardial tissue.

4.2 Autologous Source of Progenitor Cells for Perinatal and Fetal Cardiac Tissue Engineering

As most anatomic defects, including those of the heart, are routinely diagnosed during the first trimester of gestation by prenatal ultrasonography, a recently described concept of perinatal tissue engineering has emerged. This therapeutic strategy involves in utero harvest of fetal tissue, isolation, expansion in vitro, and organization of such autologous cells on a synthetic matrix creating an engineered tissue construct. As all manipulation occurs parallel to ongoing gestation, at birth the infant can benefit from surgical reconstruction utilizing this autologous tissue. To date, this concept has been explored as experimental therapeutic strategy for reconstruction of muscular diaphragmatic defects, bladder extrophy, as well as superficial skin defects (Fauza et al. 1998a,b, 2001). Due to donor site morbidity and the difficulty of in vitro expansion of differentiated cardiac myocytes, such an approach cannot be applied to myocardial tissue engineering. As numerous investigators have described cardiomyocytic differentiation potential of adult bone marrow-derived mesenchymal progenitor cells, we hypothesized that the fetal liver, the hematopoietic organ in utero, might contain a similar population of cells (Makino et al. 1999; Liechty et al. 2000).

To evaluate this possibility we modified previously described protocols for the isolation and expansion of adult bone marrow-derived mesenchymal progenitor cells and applied them to the fetal liver (Pittenger et al. 1999). For this set of experiments we chose to work in an ovine model as this animal has been utilized extensively for studies of both fetal surgery and perinatal tissue engineering with numerous established experimental models of human disease such as congenital diaphragmatic hernia, myelomeningocele, abdominal wall defects, as well as congenital cardiac defects (Fishman et al. 1978; Meuli et al. 1995; Quinn et al. 1999; Luton et al. 2000). Comparable size of the human and ovine fetus, as well as minimal uterine irritability and low rates of abortion, even after an open hysterotomy, further increase the utility of this animal as a model of fetal surgery (Krupnick and Adzick 2001).

Tissue was obtained by resecting a segment of the right lobe of the liver in an approximately 100-day gestational fetus (term=145 days). The liver was mechanically degraded, enzymatically digested and separated on a density gradient. Low-density cells were collected and enriched for mesenchymal progenitor cells by plastic adherence. The media was changed at 48 h to remove nonadherent cells and the adherent cell layer was allowed to expand with bi-weekly media changes. Passage 1 cells displayed a uniform fibroblast-like morphology, but when exposed to 24 h of 5-azacytidine or cultured with dexamethasone, this cell population fused into elongated multinucleated cells resembling skeletal mus-

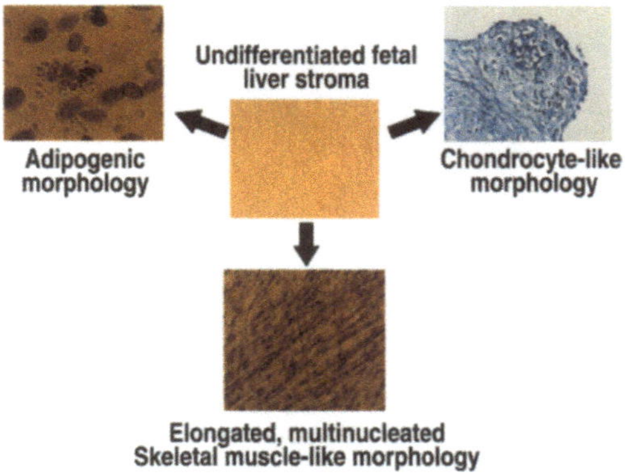

Fig. 4 In vitro differentiation of fetal liver stroma. While early passage cells had a uniform fibroblast-like morphology under proper in vitro conditions these cells could differentiate into adipocytes, chondrocytes, skeletal myocytes and spontaneously contracting cells resembling cardiac myocytes

cle. Expression of mRNA for MyoD, a skeletal muscle specific transcription factor was detected by reverse transcription–polymerase chain reaction (Davis et al. 1987) and immunohistochemical evaluation revealed expression of skeletal myosin heavy chain organized in a sarcomeric pattern (data not shown). The same cells grown in the presence of insulin, dexamethasone, isobutylmethylxanthine, and indomethacin, developed an adipocyte-like morphology exhibiting cytoplasmic lipid droplets intensely staining with Oil-Red-O and if cultured as a pellet micromass in serum-free medium, the culture formed lacunae and synthesized a proteoglycan rich extracellular matrix staining with Alcian Blue (Pittenger et al. 1999). If cultured over a bed of type IV collagen and laminin supplemented with growth factors (Matrigel) spontaneously contracting cells resembling cardiac myocytes could be generated (Fig. 4). More importantly, when organized on a three-dimensional matrix and implanted into the left ventricle of an immunocompromised rat, as described above, cardiomyocytic differentiation was identified by cytoplasmic expression of cardiomyocyte specific proteins in the transplanted cells (Fig. 5) (A.S. Krupnick, K.R. Balsara, D. Kreisel, A.E. Gelman, M. Riha, M.S. Estives, B.R. Rosengard and A.W. Flake, unpublished results). As a validation of the clinical assumption that autologous fetal liver can be harvested in utero, in a parallel set of studies our collegues at The Children's Hospital of Philadelphia have been able to perform liver resection

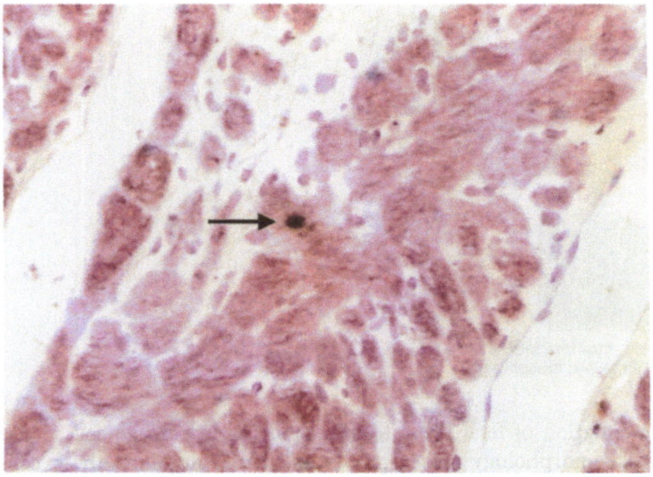

Fig. 5 In vivo transplantation of fetal liver stroma. BrdU labeled fetal liver stromal cells implanted into the left ventricle of a nude rat underwent cardiomyocytic differentiation as noted by co-localization of BrdU nuclear stain (black) with anti-cardiac troponin C cytoplasmic staining (violet)

on midgestation fetal sheep, similar to methodology utilized in this communication, with a 75% fetal survival (A.W. Flake, unpublished results). Thus the fetal liver could potentially serve as an autologous source of progenitor cells for myocardial tissue engineering in the future. Ongoing improvements in post-fetal surgery tocolysis and the rapidly expanding development of fetoscopy, including minimally invasive surgical manipulation, offers the potential to minimize mortality after in utero harvest of fetal tissue further, thus making perinatal cardiovascular tissue engineering a clinical possibility in the near future (Fauza et al. 1998a,b, 2001; Harrison et al. 1998).

5 Current Limitations and Future Directions of Cardiac Tissue Engineering

Despite encouraging experimental results, limitation of nutrient diffusion into thick three-dimensional engineered tissue remains the Achilles' heel of tissue engineering and has prevented the successful construction and transplantation of all but the simplest structures. Similar to previous studies we were able to obtain only limited engraftment and survival of cells in the area immediately adjacent to host myocardium, closest to the native blood supply, but cells in the center of the construct did not survive (Carrier et al. 2002). Numerous solutions such as incorporation of angiogenic peptides, co-culture with vascular endothelium, and angiogenic gene therapy have been attempted in order to accelerate neoangiogenesis into the engineered construct and increase cell survival (Shea et al. 1999; Nor et al. 2001; Richardson et al. 2001). Despite these efforts, such solutions maybe limited as vascular ingrowth into thick structures occurs over several days while cell death due to lack of mass nutrient transport could occur within hours. A novel therapeutic approach has recently been reported that directly address the limitation of neovascularization of engineered tissue. Rather then relying on vascular ingrowth into an engineered construct, Kaihara and colleagues have utilized micromachining technology in an attempt to engineer a vascular network (Kaihara et al. 2000). After etching a branching vascular network pattern onto the surface of silicon and Pyrex wafers, a layer of endothelial cells was placed onto the wafers and evaluated for its ability to organize into a three-dimensional structure. The investigators found that after a period of time the endothelial cells could form single-cell alignments and the vascular structure could be removed from the wafer as an intact monolayer. While preliminary in nature, such work could form the basis of solid organ engineering with the tissue of interest constructed around a prefabricated vascular network which would then be perfused by a surgical anastomosis to the recipient blood supply—similar to the current method of solid organ transplantation.

6 Myocardial Regeneration

The belief that cardiac muscle consists of fully differentiated cells, which are unable to synthesize DNA and undergo mitotic division, has its basis in early twentieth century anatomic studies (Karsner et al. 1925). For the last several decades, this line of thinking limited active investigation into myocardial regeneration. Moreover, the assumption that unlike the satellite cells present in skeletal muscle, cardiac myocytes lack a viable source of stem cells for regeneration further hindered research into this field. Recent data has all but invalidated such assumptions. Adult mammalian cardiac myocytes can synthesize DNA and reenter the cell cycle after myocardial infarction as well as after heart failure from nonischemic causes (Linzbach 1960; Kajstura et al. 1998; Beltrami et al. 2001). The recent finding that several lineages of bone marrow-derived cells are able to serve as a source of cardiac myocytes after myocardial infarction provides further evidence that adult mammalian myocardium can regenerate (Jackson et al. 2001; Orlic et al. 2001b). Since the MRL mouse strain can regenerate right ventricular myocardium after a cryogenic infarction with near complete restoration of function, it is apparent that the potential for myocardial regeneration is not lost in mammalian species (Leferovich et al. 2001). Despite these notions ischemic heart failure remains the number one cause of death in the Western world and complete regeneration of infarcted heart tissue does not occur as the majority of the infarct is replaced by scar tissue rather than newly formed cardiac muscle. Interestingly, the regenerating myocardium described in most experimental models is located at the edge of the myocardial scar rather in the center of the actual infarct. While this could be the effect of limited blood supply to the scar tissue, alteration in the extracellular matrix of myocardial scar might affect both regeneration and neoangiogenesis of the infarcted muscle. As the extracellular matrix of the peri-infarct area is generally preserved, while the matrix in the center of the myocardial scar is destroyed as a by-product of myocyte debridement by invading leukocytes, it is a theoretical possibility that the environmental cues necessary for regeneration are lost from the cardiac matrix during this debridement process. If this were the case it might be possible to boost regeneration by replacing the extracellular matrix of an infarct with that of healthy, normal myocardium.

Similar to myocardium, injury to the skin resulting from a full-thickness burn is repaired through the formation of scar tissue rather than regeneration. To address the clinical need for skin replacement investigators at LifeCell Corporation (Branchburg, NJ, USA) have developed novel protocols to remove the epidermis and cellular elements of the dermis from allogeneic skin without altering the structure of the extracellular matrix (Livesey et al. 1995). When utilized as a graft to replace full-thickness injury to the dermis and epidermis, this matrix supports regeneration of the dermal matrix with repopulation by fibroblasts, blood vessels and reconstitution of the epidermis after coverage by a split thickness skin graft (Wainwright et al. 1996; Reagan et al. 1997). This is accomplished with a minimal inflammatory cell infiltrate and without tissue rejection. To date this form of therapy termed 'guided tissue regeneration' utilizing processed naturally derived matrix from various sites, such as skin (Alloderm, LifeCell Corporation, Branchburg, NJ) or small intestine submucosa (Surgisis, Cook Biotechnology, West Lafayette, IN, USA) has been utilized to facilitate regeneration of dermal tissue as well as the esophagus, small intestine, and bladder (Badylak et al. 1998, 2000; Chen and Badylak 2001).

As the ability to harness this form of matrix therapy and apply it to myocardial regeneration offers numerous advantages over currently available therapeutic strategies of myocardial replacement, we set out to evaluate the ability of acellular myocardial matrix to support ventricular tissue regeneration (Krupnick et al. 2002b). Utilizing protocols modified from decellularization of vascular tissue, myocardium was obtained from adult rats, processed to remove cellular elements and then freeze-dried. Within a month of processing the matrix was rehydrated and evaluated histologically and immunohistochemically for architectural integrity and the presence of contractile proteins. Overall the matrix retained ultrastructural characteristics of normal myocardium with conspicuous absence of visible nuclei. Immunohistochemical evaluation revealed remnants of contractile proteins such as α-actinin, myosin heavy chain, and connexin 43 (Fig. 6).

This matrix was then used to replace full thickness myocardial defects created after excision of a cryogenic infarct in the left ventricle (Fig. 7, left and middle panels). The repaired heart was transplanted heterotopically into a syngeneic animal in order to create a fully loaded left ventricle, as described above, and the rats were given continuous BrdU in their drinking water in order to label dividing cells (Fig. 7, right panel). Cryogenic infarctions without excision and matrix replacement were created in a parallel set of animals and served as a negative control. Animals from both groups were sacrificed at bi-weekly intervals, up to 2 month post-transplantation, in order to evaluate regeneration. After 2 weeks minimal resorption of the matrix had occurred with evidence of moderate inflammation and neoangiogenesis in the interface between the matrix and native myocardium (Fig. 8). Complete resorption of the matrix was evident by 8 weeks after transplantation with replacement of the construct by host tissue. Regenerating myocytes were evaluated in the peri-infarct and peri-matrix region by staining for both nuclear BrdU and cytoplasmic myosin heavy chain. DNA containing BrdU was evident in 0.59±0.1% of the cardiac myocytes in the peri-infarct region, within 200 μm of the cryogenic infarct, in the control group which was statisti-

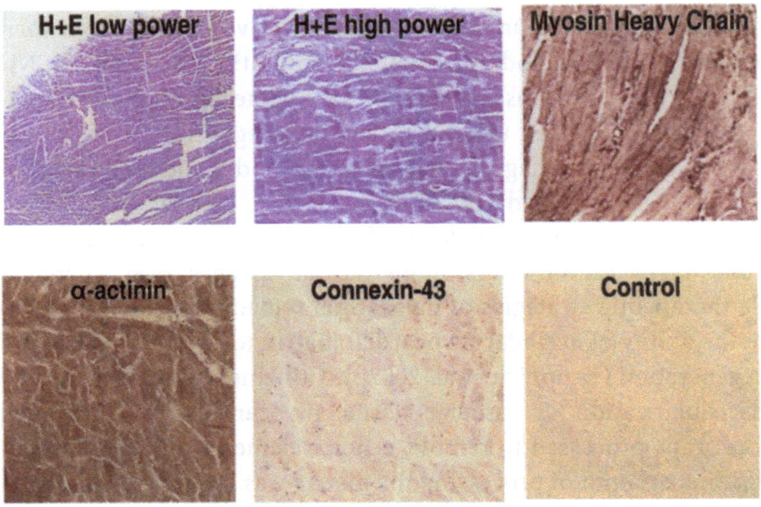

Fig. 6 Processed cardiac matrix retained the ultrastructural characteristics of normal myocardium on hematoxylin and eosin staining with notable absence of nuclear staining. Immunohistochemical evaluation revealed that the cellular remnants contained contractile proteins such as myosin heavy chain and α-actinin as well as gap the junctional protein connexin 43. No background was evident when staining with a nonspecific mouse isotype control

cally identical to the number of BrdU positive myocytes in the experimental matrix replacement group 0.55±0.06% (P=0.78). Despite these initially discouraging results, this is the first reported attempt of utilizing cryopreserved matrix to guide myocardial regeneration.

There are several potential barriers that we encountered using this experimental approach. Unlike other organs such as skin and even bowel, the viable and well perfused heart undergoes continuous cyclic and forceful contractions. As these contractions create ongoing motion between the native myocardial tissue and the static transplanted matrix, disruption of vascular ingrowth and limitation of angiogenesis may occur initially and could affect the regenerative response. Furthermore, as the heart is a highly metabolic organ with an extensive blood supply, postoperative bleeding and the development of thrombus at the edges of the implanted matrix is difficult to avoid. This thrombus may act as a repository for proinflammatory cytokines setting up an inflammatory nidus and leading to eventual destruction of the transplanted matrix rather than regeneration.

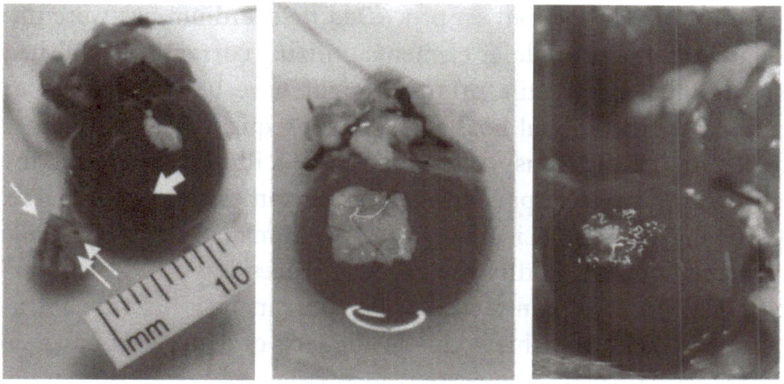

Fig. 7 *Left*: The myocardial matrix (*arrow*) reinforced by a piece of autologous aorta (*double arrow*) was used to replaced a full thickness myocardial defect created after the excision of a cryogenic infarct (*arrowhead*). *Middle*: Complete apposition to the native myocardium could be achieved by a running 9-0 nylon suture placed through the aortic reinforcement and epicardial surface of the heart. *Right*: Reperfusion of this heart resulted in some bleeding at the matrix/myocardial interface which was controlled by gentle pressure on the matrix leading to localized clot formation

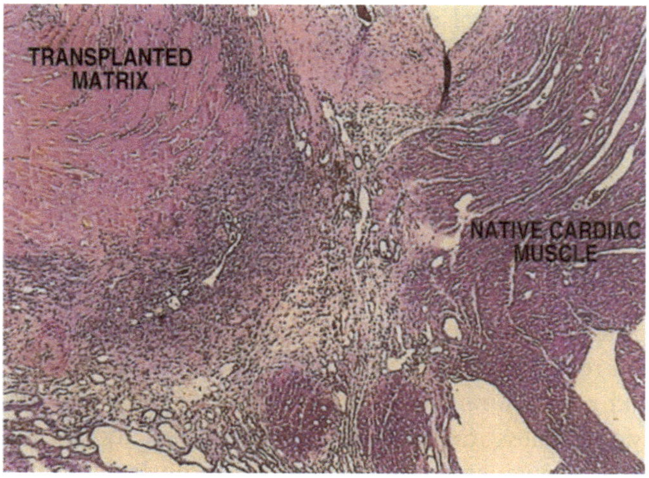

Fig. 8 As early as 2 weeks after transplantation some resorption of the cryopreserved matrix was noted with ingrowth of blood vessels and a moderate inflammatory response

While we intuitively focused on processed myocardium, we were unable to remove all the subcellular elements utilizing currently established protocols. Residual contractile proteins might have been perceived as necrotic tissue by the animal host with the subsequential inflammatory response destroying the transplanted matrix. It is important to note the lack of an inflammatory response which accompanies successful matrix guided tissue regeneration (Livesey et al. 1995; Allman et al. 2001). While it is our ongoing goal to modify the existing processing protocols in order to eliminate residual contractile proteins in myocardial tissue, this goal might be difficult to achieve due to the lack of a compact collagen matrix in the heart, such as is present in the dermis. As a structurally rigorous matrix is necessary to support tissue integrity after decellularization, protocols geared toward complete cell removal might destroy the extracellular structure of the fragile cardiac matrix. Since tissue regeneration could simply depend on the presence of an undamaged collagenous matrix placed within the proper environment, it is possible that other acellular tissue could support myocardial regeneration when placed properly into heart. This notion is supported by the ability of dermal matrix to guide regeneration of cartilage, dura, and gingival tissue under proper environmental conditions (Harris 1998; Kridel et al. 1998; Chaplin et al. 1999). These as well as other novel therapeutic applications are currently being investigated.

7
Conclusions

As heart disease remains a major source of morbidity and mortality in the adult and pediatric population, an ever growing demand exists for innovative therapy aimed at treating these diseases. While allogeneic organ transplantation remains a viable therapeutic option for some, donor organ shortage and rejection remain formidable problems for most patients. The ability to harness myocardial regenerative capacity and utilize principles of tissue engineering for myocardial reconstruction may offer novel therapeutic possibilities to combat cardiac disease.

References

Allman AJ, McPherson TB, Badylak SF, Merrill LC, Kallakury B, Sheehan C, Raeder RH, Metzger DW (2001) Xenogeneic extracellular matrix grafts elicit a TH2-restricted immune response. Transplantation 71:1631–1640

Atkins BZ, Hueman MT, Meuchel JM, Cottman MJ, Hutcheson KA, Taylor DA (1999) Myogenic cell transplantation improves in vivo regional performance in infarcted rabbit myocardium. Journal of Heart & Lung Transplantation 18:1173–1180

Badylak S, Meurling S, Chen M, Spievack A, Simmons-Byrd A (2000) Resorbable bioscaffold for esophageal repair in a dog model. Journal of Pediatric Surgery 35:1097–1103

Badylak SF, Kropp B, McPherson T, Liang H, Snyder PW (1998) Small intestional submucosa: a rapidly resorbed bioscaffold for augmentation cystoplasty in a dog model. Tissue Engineering 4:379–387

Beltrami AP, Urbanek K, Kajstura J, Yan SM, Finato N, Bussani R, Nadal-Ginard B, Silvestri F, Leri A, Beltrami AC, Anversa P (2001) Evidence that human cardiac myocytes divide after myocardial infarction. New England Journal of Medicine 344:1750–1757

Benichou G, Valujskikh A, Heeger PS (1999) Contributions of direct and indirect T cell alloreactivity during allograft rejection in mice. Journal of Immunology 162:352–358

Carrier RL, Papadaki M, Rupnick M, Schoen FJ, Bursac N, Langer R, Freed LE, Vunjak-Novakovic G (1999) Cardiac tissue engineering: cell seeding, cultivation parameters, and tissue construct characterization. Biotechnology & Bioengineering 64:580–589

Carrier RL, Rupnick M, Langer R, Schoen FJ, Freed LE, Vunjak-Novakovic G (2002) Perfusion improves tissue architecture of engineered cardiac muscle. Tissue Engineering 8:175–188

Chaplin JM, Costantino PD, Wolpoe ME, Bederson JB, Griffey ES, Zhang WX (1999) Use of an acellular dermal allograft for dural replacement: an experimental study. Neurosurgery 45:320–327

Chen MK, Badylak SF (2001) Small bowel tissue engineering using small intestinal submucosa as a scaffold. Journal of Surgical Research 99:352–358

Condorelli G, Borello U, De Angelis L, Latronico M, Sirabella D, Coletta M, Galli R, Balconi G, Follenzi A, Frati G, Cusella De Angelis MG, Gioglio L, Amuchastegui S, Adorini L, Naldini L, Vescovi A, Dejana E, Cossu G (2001) Cardiomyocytes induce endothelial cells to trans-differentiate into cardiac muscle: implications for myocardium regeneration. Proceedings of the National Academy of Sciences of the United States of America 98:10733–10738

Davis RL, Weintraub H, Lassar AB (1987) Expression of a single transfected cDNA converts fibroblasts to myoblasts. Cell 51:987–1000

Ensminger SM, Spriewald BM, Witzke O, Pajaro OE, Yacoub MH, Morris PJ, Rose ML, Wood KJ (2002) Indirect allorecognition can play an important role in the development of transplant arteriosclerosis. Transplantation 73:279–286

Fauza DO, Fishman SJ, Mehegan K, Atala A (1998a) Videofetoscopically assisted fetal tissue engineering: bladder augmentation. Journal of Pediatric Surgery 33:7–12

Fauza DO, Fishman SJ, Mehegan K, Atala A (1998b) Videofetoscopically assisted fetal tissue engineering: skin replacement. Journal of Pediatric Surgery 33:357–361
Fauza DO, Marler JJ, Koka R, Forse RA, Mayer JE, Vacanti JP (2001) Fetal tissue engineering: diaphragmatic replacement. Journal of Pediatric Surgery 36:146–151
Fishman NH, Hof RB, Rudolph AM,Heymann MA (1978) Models of congenital heart disease in fetal lambs. Circulation 58:354–364
Harris RJ (1998) Root coverage with a connective tissue with partial thickness double pedicle graft and an acellular dermal matrix graft: a clinical and histological evaluation of a case report. Journal of Periodontology 69:1305–1311
Harrison MR, Mychaliska GB, Albanese CT, Jennings RW, Farrell JA, Hawgood S, Sandberg P, Levine AH, Lobo E, Filly RA (1998) Correction of congenital diaphragmatic hernia in utero IX: fetuses with poor prognosis (liver herniation and low lung-to-head ratio) can be saved by fetoscopic temporary tracheal occlusion. Journal of Pediatric Surgery 33:1017–1022; discussion 1022–1013
Hoerstrup SP, Sodian R, Sperling JS, Vacanti JP, Mayer JE, Jr. (2000) New pulsatile bioreactor for in vitro formation of tissue engineered heart valves. Tissue Engineering 6:75–79
Hollenberg SM (2001) Coronary endothelial dysfunction after heart transplantation predicts allograft vasculopathy and cardiac death. Circulation 104:3091–3096
Hutcheson KA, Atkins BZ, Hueman MT, Hopkins MB, Glower DD, Taylor DA (2000) Comparison of benefits on myocardial performance of cellular cardiomyoplasty with skeletal myoblasts and fibroblasts. Cell Transplantation 9:359–368
Jackson KA, Majka SM, Wang H, Pocius J, Hartley CJ, Majesky MW, Entman ML, Michael LH, Hirschi KK, Goodell MA (2001) Regeneration of ischemic cardiac muscle and vascular endothelium by adult stem cells. Journal of Clinical Investigation 107:1395–1402
Kaihara S, Borenstein J, Koka R, Lalan S, Ochoa ER, Ravens M, Pien H, Cunningham B,Vacanti JP (2000) Silicon micromachining to tissue engineer branched vascular channels for liver fabrication. Tissue Engineering 6:105–117
Kajstura J, Leri A, Finato N, Di Loreto C, Beltrami AC, Anversa P (1998) Myocyte proliferation in end-stage cardiac failure in humans. Proceedings of the National Academy of Sciences of the United States of America 95:8801–8805
Karsner HT, Saphir O, Todd TW (1925) The State of Cradiac Muscle in Hypertrophy and Atrophy. American Journal of Pathology 1:351–371
Kelley ST (1999) Restraining infarct expansion preserves left ventricular geometry and function after acute anteroapical infarction. Circulation 99:135–142
Kreisel D, Krupnick AS, Gelman AE, Engels FH, Popma SH, Krasinskas AM, Balsara KR, Szeto WY, Turka LA, Rosengard BR (2002) Non-hematopoietic allograft cells directly activate CD8+ T cells and trigger acute rejection: an alternative mechanism of allorecognition. Nature Medicine 8:233–239
Kridel RW, Foda H, Lunde KC (1998) Septal perforation repair with acellular human dermal allograft. Archives of Otolaryngology – Head & Neck Surgery 124:73–78
Krupnick A, Adzick NS (2001) Research in fetal surgery. In: Souba WW and Wilmore DW (eds) Surgical Research. Academic Press, New York, pp 1065–1079
Krupnick A, Kreisel D, Engels F, Szeto W, Plappert T, Popma SH, Flake AW, Rosengard BR (2002a) A novel small animal model of ventricular tissue engineering. The Journal of Heart & Lung Transplantation 21:233–243

Krupnick AS, Balsara KR, Kreisel D, Gelman AE, Riha M, Estives MS, Rosengard BR, Flake AW (2003) The fetal liver as a source of multilineage progenitor cells: implications for fetal tissue engineering. submitted for publication

Krupnick AS, Kreisel D, Riha M, Balsara KR, Boerboom LE, Livesey SA, Rosengard BR (2002b) A small animal model to study ventricular regeneration. American Journal of Transplantation 2:222

Langer R,Vacanti JP (1993) Tissue engineering. Science 260:920–926

Lee RS, Yamada K, Houser SL, Womer KL, Maloney ME, Rose HS, Sayegh MH, Madsen JC (2001) Indirect recognition of allopeptides promotes the development of cardiac allograft vasculopathy. Proceedings of the National Academy of Sciences of the United States of America 98:3276–3281

Leferovich JM, Bedelbaeva K, Samulewicz S, Zhang XM, Zwas D, Lankford EB, Heber-Katz E (2001) Heart regeneration in adult MRL mice. Proceedings of the National Academy of Sciences of the United States of America 98:9830–9835

Leor J, Patterson M, Quinones MJ, Kedes LH, Kloner RA (1996) Transplantation of fetal myocardial tissue into the infarcted myocardium of rat. A potential method for repair of infarcted myocardium? Circulation 94:II332–336

Li RK (1999) Smooth muscle cell transplantation into myocardial scar tissue improves heart function. Journal of Molecular & Cellular Cardiology 31:513–522

Li RK, Jia ZQ, Weisel RD, Mickle DA, Zhang J, Mohabeer MK, Rao V, Ivanov J (1996a) Cardiomyocyte transplantation improves heart function. Annals of Thoracic Surgery 62:654–660; discussion 660–651

Li RK, Mickle DA, Weisel RD, Mohabeer MK, Zhang J, Rao V, Li G, Merante F, Jia ZQ (1997) Natural history of fetal rat cardiomyocytes transplanted into adult rat myocardial scar tissue. Circulation 96:II-179–186; discussion 186–177

Li RK, Mickle DA, Weisel RD, Zhang J, Mohabeer MK (1996b) In vivo survival and function of transplanted rat cardiomyocytes. Circulation Research 78:283–288

Li RK, Weisel RD, Mickle DA, Jia ZQ, Kim EJ, Sakai T, Tomita S, Schwartz L, Iwanochko M, Husain M, Cusimano RJ, Burns RJ, Yau TM (2000) Autologous porcine heart cell transplantation improved heart function after a myocardial infarction. Journal of Thoracic & Cardiovascular Surgery 119:62–68

Liechty KW, MacKenzie TC, Shaaban AF, Radu A, Moseley AM, Deans R, Marshak DR, Flake AW (2000) Human mesenchymal stem cells engraft and demonstrate site-specific differentiation after in utero transplantation in sheep. Nature Medicine 6:1282–1286

Linzbach AJ (1960) Heart Failure from the Point of View of Quantitative Anatomy. American Journal of Cardiology 5:370–382

Liu YH, Yang XP, Nass O, Sabbah HN, Peterson E, Carretero OA (1997) Chronic heart failure induced by coronary artery ligation in Lewis inbred rats. American Journal of Physiology 272:H722–727

Livesey SA, Herndon DN, Hollyoak MA, Atkinson YH, Nag A (1995) Transplanted acellular allograft dermal matrix. Potential as a template for the reconstruction of viable dermis. Transplantation 60:1–9

Luton D, de Lagausie P, Guibourdenche J, Peuchmaur M, Sibony O, Aigrain Y, Oury JF, Blot P (2000) Influence of amnioinfusion in a model of in utero created gastroschisis in the pregnant ewe. Fetal Diagnosis & Therapy 15:224–228

Makino S, Fukuda K, Miyoshi S, Konishi F, Kodama H, Pan J, Sano M, Takahashi T, Hori S, Abe H, Hata J, Umezawa A, Ogawa S (1999) Cardiomyocytes can be generated from marrow stromal cells in vitro. Journal of Clinical Investigation 103:697–705

Menasche P, Hagege AA, Scorsin M, Pouzet B, Desnos M, Duboc D, Schwartz K, Vilquin JT, Marolleau JP (2001) Myoblast transplantation for heart failure. Lancet 357:279–280

Meuli M, Meuli-Simmen C, Hutchins GM, Yingling CD, Hoffman KM, Harrison MR, Adzick NS (1995) In utero surgery rescues neurological function at birth in sheep with spina bifida. Nature Medicine 1:342–347

Morshead CM, Benveniste P, Iscove NN, van der Kooy D (2002) Hematopoietic competence is a rare property of neural stem cells that may depend on genetic and epigenetic alterations. Nature Medicine 8:268-273

Niklason LE, Gao J, Abbott WM, Hirschi KK, Houser S, Marini R, Langer R (1999) Functional arteries grown in vitro. Science 284:489–493

Nor JE, Peters MC, Christensen JB, Sutorik MM, Linn S, Khan MK, Addison CL, Mooney DJ, Polverini PJ (2001) Engineering and characterization of functional human microvessels in immunodeficient mice. Laboratory Investigation 81:453–463

Ono K, Lindsey ES (1969) Improved technique of heart transplantation in rats. Journal of Thoracic & Cardiovascular Surgery 57:225–229

Orlic D, Kajstura J, Chimenti S, Jakoniuk I, Anderson SM, Li B, Pickel J, McKay R, Nadal-Ginard B, Bodine DM, Leri A, Anversa P (2001a) Bone marrow cells regenerate infarcted myocardium. Nature 410:701–705

Orlic D, Kajstura J, Chimenti S, Limana F, Jakoniuk I, Quaini F, Nadal-Ginard B, Bodine DM, Leri A, Anversa P (2001b) Mobilized bone marrow cells repair the infarcted heart, improving function and survival. Proceedings of the National Academy of Sciences of the United States of America 98:10344–10349

Pfeffer JM (1991) Progressive ventricular remodeling in rat with myocardial infarction. American Journal of Physiology 260:H1406–1414

Pittenger MF, Mackay AM, Beck SC, Jaiswal RK, Douglas R, Mosca JD, Moorman MA, Simonetti DW, Craig S, Marshak DR (1999) Multilineage potential of adult human mesenchymal stem cells. Science 284:143–147

Pouzet B (2000) Intramyocardial transplantation of autologous myoblasts: can tissue processing be optimized? Circulation 102:III210–215

Quinn TM, Sylvester KG, Kitano Y, Liechty KW, Jarrett BP, Adzick NS, Flake AW (1999) TGF-beta2 is increased after fetal tracheal occlusion. Journal of Pediatric Surgery 34:701–704; discussion 704–705

Rajnoch C, Chachques JC, Berrebi A, Bruneval P, Benoit MO, Carpentier A (2001) Cellular therapy reverses myocardial dysfunction. Journal of Thoracic & Cardiovascular Surgery 121:871–878

Raper SE, Yudkoff M, Chirmule N, Gao GP, Nunes F, Haskal ZJ, Furth EE, Propert KJ, Robinson MB, Magosin S, Simoes H, Speicher L, Hughes J, Tazelaar J, Wivel NA, Wilson JM, Batshaw ML (2002) A pilot study of in vivo liver-directed gene transfer with an adenoviral vector in partial ornithine transcarbamylase deficiency. Human Gene Therapy 13:163–175

Reagan BJ, Madden MR, Huo J, Mathwich M, Staiano-Coico L (1997) Analysis of cellular and decellular allogeneic dermal grafts for the treatment of full-thickness wounds in a porcine model. Journal of Trauma-Injury Infection & Critical Care 43:458–466

Reinecke H, Zhang M, Bartosek T, Murry CE (1999) Survival, integration, and differentiation of cardiomyocyte grafts: a study in normal and injured rat hearts. Circulation 100:193–202

Richardson TP, Peters MC, Ennett AB, Mooney DJ (2001) Polymeric system for dual growth factor delivery. [see comments.]. Nature Biotechnology 19:1029–1034

Sakai T (1999) Fetal cell transplantation: a comparison of three cell types. Journal of Thoracic & Cardiovascular Surgery 118:715–724

Sakai T, Li RK, Weisel RD, Mickle DA, Kim EJ, Tomita S, Jia ZQ, Yau TM (1999) Autologous heart cell transplantation improves cardiac function after myocardial injury. Annals of Thoracic Surgery 68:2074–2080; discussion 2080–2071

Scorsin M, Hagege AA, Marotte F, Mirochnik N, Copin H, Barnoux M, Sabri A, Samuel JL, Rappaport L, Menasche P (1997) Does transplantation of cardiomyocytes improve function of infarcted myocardium? Circulation 96:II-188–193

Scorsin M, Marotte F, Sabri A, Le Dref O, Demirag M, Samuel JL, Rappaport L, Menasche P (1996) Can grafted cardiomyocytes colonize peri-infarct myocardial areas? Circulation 94:II337–340

Shea LD, Smiley E, Bonadio J, Mooney DJ (1999) DNA delivery from polymer matrices for tissue engineering. [see comments.] [erratum appears in Nat Biotechnol 1999 Aug;17(8):817.]. Nature Biotechnology 17:551–554

Shinoka T, Breuer CK, Tanel RE, Zund G, Miura T, Ma PX, Langer R, Vacanti JP, Mayer JE, Jr. (1995) Tissue engineering heart valves: valve leaflet replacement study in a lamb model. Annals of Thoracic Surgery 60:S513–516

Shinoka T, Ma PX, Shum-Tim D, Breuer CK, Cusick RA, Zund G, Langer R, Vacanti JP, Mayer JE, Jr. (1996) Tissue-engineered heart valves. Autologous valve leaflet replacement study in a lamb model. Circulation 94:II164–168

Snabaitis AK, Shattock MJ, Chambers DJ (1997) Comparison of polarized and depolarized arrest in the isolated rat heart for long-term preservation. Circulation 96:3148–3156

Soonpaa MH, Koh GY, Klug MG, Field LJ (1994) Formation of nascent intercalated disks between grafted fetal cardiomyocytes and host myocardium. Science 264:98–101

Srivastava D, Olson EN (2000) A genetic blueprint for cardiac development. Nature 407:221–226

Suchin EJ, Langmuir PB, Palmer E, Sayegh MH, Wells AD, Turka LA (2001) Quantifying the frequency of alloreactive T cells in vivo: new answers to an old question. Journal of Immunology 166:973–981

Taylor DA (1998) Regenerating functional myocardium: improved performance after skeletal myoblast transplantation. Nature Medicine 4:929–933

Toma C, Pittenger MF, Cahill KS, Byrne BJ, Kessler PD (2002) Human mesenchymal stem cells differentiate to a cardiomyocyte phenotype in the adult murine heart. Circulation 105:93–98

Tomita S, Li RK, Weisel RD, Mickle DA, Kim EJ, Sakai T, Jia ZQ (1999) Autologous transplantation of bone marrow cells improves damaged heart function. Circulation 100:II247–256

Tremblay JP, Guerette B (1997) Myoblast transplantation: a brief review of the problems and of some solutions. Basic and Applied Myology 7:221–230

Tsuyuki RT (2001) Acute precipitants of congestive heart failure exacerbations. Archives of Internal Medicine 161:2337–2342

Vacanti JP, Langer R (1999) Tissue engineering: the design and fabrication of living replacement devices for surgical reconstruction and transplantation. Lancet 354 Suppl 1:SI32–34

Wainwright D, Madden M, Luterman A, Hunt J, Monafo W, Heimbach D, Kagan R, Sittig K, Dimick A, Herndon D (1996) Clinical evaluation of an acellular allograft dermal matrix in full-thickness burns. Journal of Burn Care & Rehabilitation 17:124–136

Yokomuro H, Li RK, Mickle DA, Weisel RD, Verma S, Yau TM (2001) Transplantation of cryopreserved cardiomyocytes. Journal of Thoracic & Cardiovascular Surgery 121:98–107

Yoo KJ, Li RK, Weisel RD, Mickle DA, Jia ZQ, Kim EJ, Tomita S, Yau TM (2000) Heart cell transplantation improves heart function in dilated cardiomyopathic hamsters. Circulation 102:III204–209

Zimmermann WH, Fink C, Kralisch D, Remmers U, Weil J, Eschenhagen T (2000) Three-dimensional engineered heart tissue from neonatal rat cardiac myocytes. Biotechnology & Bioengineering 68:106–114

Spallanzani's Mouse: A Model of Restoration and Regeneration

E. Heber-Katz · J. M. Leferovich · K. Bedelbaeva · D. Gourevitch
The Wistar Institute, 3602 Spruce Street, Philadelphia, PA 19104, USA
E-mail: heberkatz@wistar.upenn.edu

1 Introduction 166
2 A Model of Epimorphic Regeneration 170
2.1 The Role of the Extracellular Matrix, the Basement Membrane, and Scar Tissue and Its Breakdown 170
2.2 The Blastema 173
3 A Model of Cardiac Regeneration 174
3.1 Cardiomyocyte Proliferation 174
3.2 The Role of Stem Cells 177
3.3 Extracellular Matrix Breakdown 178
3.4 The Angiogenic Response 179
4 The Role of Scar Tissue in Other Models of Regeneration 180
5 The Genetics of Regeneration 180
6 Conclusion 181
References 183

Abstract The ability to regenerate is thought to be a lost phenotype in mammals, though there are certainly sporadic examples of mammalian regeneration. Our laboratory has identified a strain of mouse, the MRL mouse, which has a unique capacity to heal complex tissue in an epimorphic fashion, i.e., to restore a damaged limb or organ to its normal structure and function. Initial studies using through-and-through ear punches showed rapid full closure of the ear holes with cartilage growth, new hair follicles, and normal tissue architecture reminiscent of regeneration seen in amphibians as opposed to the scarring usually seen in mammals. Since the ear hole closure phenotype is a quantitative trait, this has been used to show—through extensive breeding and backcrossing—that the trait is heritable. Such analysis reveals that there is a complex genetic basis for this trait with multiple loci. One of the major phenotypes of the MRL mouse is a potent remodeling response with the absence or a reduced level of scarring. MRL healing is associated with the upregulation of the metalloproteinases MMP-2 and MMP-9 and the downregulation of their inhibitors TIMP-2 and TIMP-3, both present in inflammatory cells such as neutrophils and macrophages. This model has more recently been extended to the heart. In this case, a cryoinjury to the right ventricle leads to near complete scarless healing in the MRL mouse whereas scarring is seen in the control mouse. In the MRL heart, bromodeoxyuridine uptake by cardiomyocytes filling the wound site can be seen 60 days after injury. This does not occur in the control mouse. Function in the MRL heart, as measured by echocardiography, returns to normal.

1 Introduction

There is a long history of observations showing that 'lower' species have the capacity to regenerate large functional units of tissue and often full limbs. Perhaps the most interesting were those made by Lazzaro Spallanzani (1729–1799), whose many contributions to science included the disproof of spontaneous generation and the first systematic study of regeneration.

He showed that the removal of a snail's head led to regrowth and regeneration of that head. This intriguing result was reproduced by many, including Voltaire. Besides snails, Spallanzani amputated salamander tails and limbs with the same striking results, complete regrowth (Dinsmore 1991). We do not know whether Spallanzani ever attempted his resection experiments in mammals because results were seldom reported.

It is generally accepted that this type of healing—the ability to regrow a limb that is perfectly reformed and functional and known as 'epimorphic' regeneration—is not seen in mammals. There have been rare exceptions and they include antler regrowth in deer (Goss 1970), mouse digit and human fingertip growth in selected cases (Douglas, 1972; Illingworth, 1974; Borgens et al. 1982; Neufield, 1989, Reginelli et al. 1995), and ear hole closure in rabbits (Goss and Grimes 1975). Thus, while lower species have provided robust, diverse, and easily accessible systems to study regeneration, mammalian examples were limited by the infrequency and/or the inefficacy of the response. Until recently, it

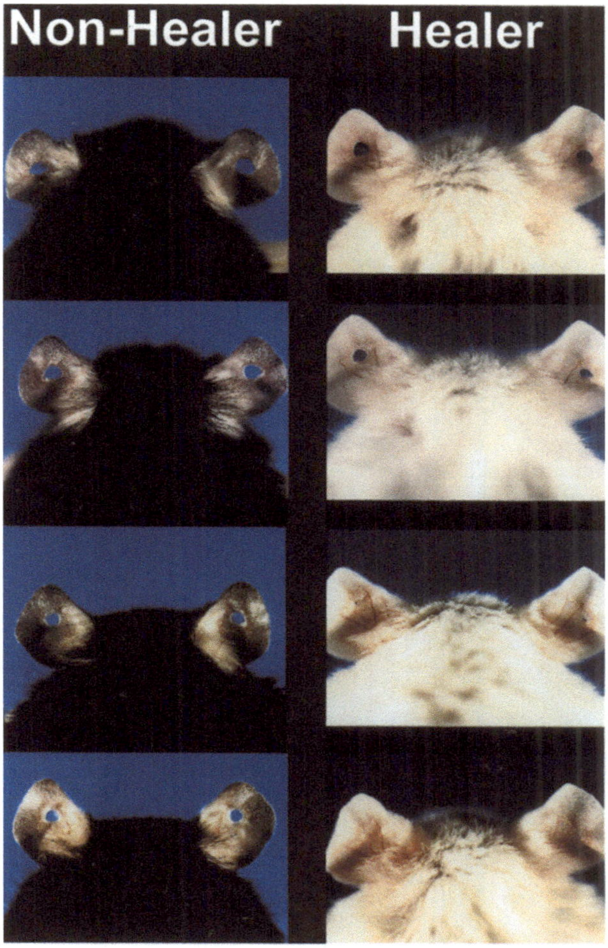

Fig. 1 The closure of ear holes in the MRL mouse over a 1-month period. C57Bl/6 nonregenerating (*left side*) and MRL/lpr regenerating (*right side*) ears were punched in the center of the ears creating a 2-mm through-and-through wound. *From top to bottom*, one can see the progression of hole closure from day 1 to day 9 to day 20 to day 33. Ear hole closure is the exception to the rule in mouse ears. (From Clark et al. 1998)

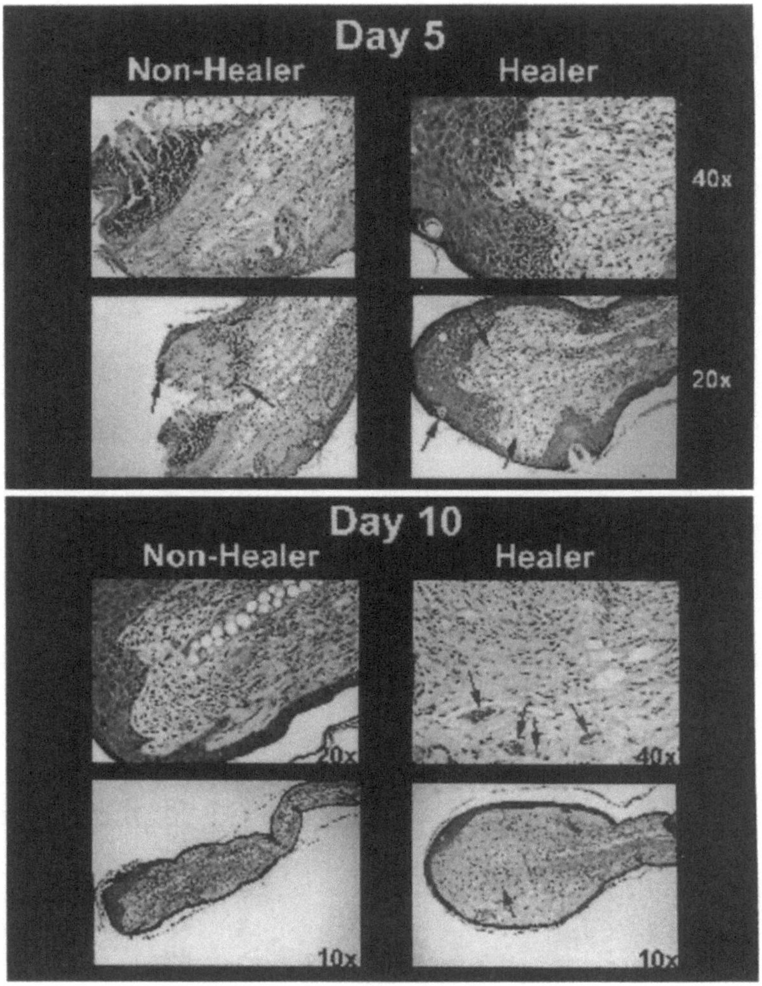

Fig. 2 The beginnings of blastema formation in the MRL mouse ear. Histological sections were cut through the ear hole various times after the holes were made. In this figure, sections from day 5 (*top*) and day 10 (*bottom*) after wounding of C57Bl/6 nonregenerating ears and MRL/lpr regenerating ears can be seen. All sections are stained with hemotoxylin and eosin. By day 5, the C57Bl/6 wound has failed to epithelialize (*arrows*) and little activity in the dermis can be seen. However, the MRL has re-epithelialized, the epidermis is quite thick, and the dermal cells appear to have proliferated (*arrows*). Also, there are new hair follicles (*arrow*) in the growth area in the MRL. By day 10, the C57BL/6 wound has completely re-epithelialized and the new epidermis is thicker than that seen in the uninjured regions of the ear, but

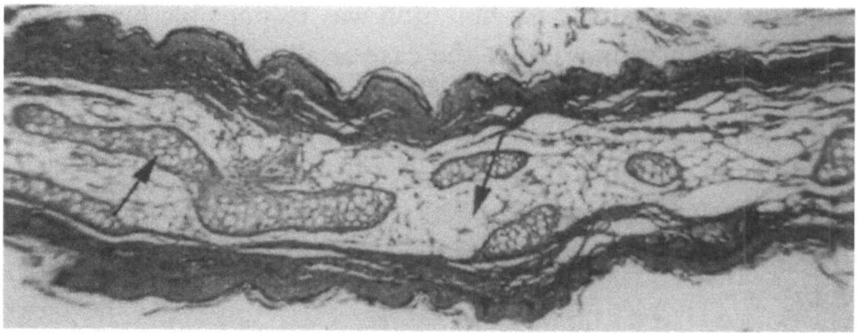

Fig. 3 Islands of cartilage appear in the closed MRL mouse ear hole. A histological section through the region of the MRL/lpr healed ear in which the ear hole was made 105 days before. Collagenous islands (*arrows*) present throughout the MRL section in which the hole has closed appear to be reorganizing. The tissue sections have been stained with Gomori trichrome stain. (From Clark et al. 1998)

was generally concluded that regeneration was simply not within the biological capacity of mammals.

Several years ago, our laboratory found that a particular mouse strain, the MRL mouse, better known for its spontaneous development of lupus-like symptoms and sequelae (Murphy and Roths 1979; Cohen and Eisenberg 1991; Theofilopoulos 1993) due to a mutation in the *fas* gene (Watanabe-Fukinaga et al. 1992; Watson et al. 1992; Adachi et al. 1993) showed unusual healing properties. We discovered that MRL mice with the *fas* mutation could close ear holes and heal completely without scarring (Fig. 1) (Clark et al. 1998; Heber-Katz, 1999). Furthermore, the healing took place with an apparent replacement of hair follicles and sebaceous glands (Fig. 2), and over time new cartilage was replaced as well (Fig. 3). This was a surprising finding because ear hole punches are commonly used as a numbering code in mice and last the lifetime of the mouse (Dickie 1975). Such healing mimicked the epimorphic regeneration reported in rabbit ears (Goss and Grimes 1975). Although our ini-

the dermis does not show signs of cell division or blastema formation. In contrast, in the MRL there is marked neovascularization (*arrows*), continued dermal cell proliferation (*arrow*) into the structure called the blastema, and the maturation of sebaceous glands in the MRL wound extending out beyond the borders of the wound where the cut cartilage edges are seen. (From Clark et al. 1998)

tial expectations were that the *fas* mutation was responsible for this effect, it turned out that this was not the case at all—MRL mice without the *fas* mutation could heal similarly. The healing trait can be traced back to a progenitor strain of MRL, the LG/J mouse, which also proved to be a healer mouse (Kench et al. 1999; E. Heber-Katz, unpublished results). Because MRL mice have inherited a great majority of their alleles from LG/J, it was thus likely that the healing trait had a sizeable heritable component.

Analysis of these mice at the genetic, molecular, and in vivo levels has both supported the original studies carried out in the amphibian and has also indicated new directions of study. We will review these data and also present several different, though not all, organ systems in this mouse that show a regenerative process.

2
A Model of Epimorphic Regeneration

The ability to heal wounds with the growth of a blastema and full replacement of tissue, complete recovery of its previous function and architecture is usually associated with amphibian limb regeneration (Stocum 1984, 1996; Tabin 1991; Gardiner and Bryant 1996; Gross 1996; Brockes 1997) as well as nonvertebrate regeneration including planaria (Sawada et al. 1999; Newmark and Sanchez Alvarado 2002), hydra (Technau et al. 2000) and sea cucumbers (Quinones et al. 2002), to name just a few. This response is very different from what is seen in mammalian wound repair (Clark 1996). However, the MRL ear hole heals in a manner that is a hybrid between mammalian wound repair and amphibian regeneration.

2.1
The Role of the Extracellular Matrix, the Basement Membrane, and Scar Tissue and Its Breakdown

One of the key events after injury is re-epithelialization of the wound. Regeneration-specific re-epithelialization is quite rapid, often taking less than a day. This is seen in many regeneration systems such as tadpole tails and cut planarial parts, and is also seen in the MRL ear hole wound. Thus, MRL re-epithelialization takes place within 2 days while the C57BL/6 takes at least 3 days and often up to 10 days (Fig. 2). A second

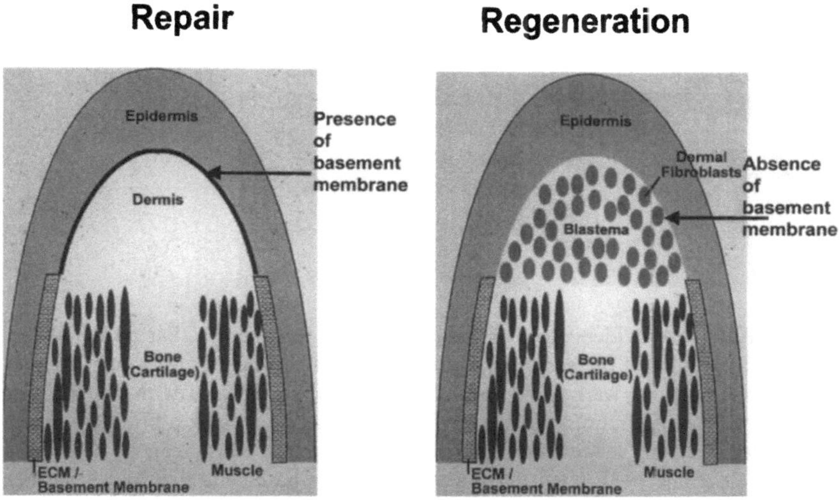

Fig. 4 Breakdown of the basement membrane and blastema formation. A cartoon comparing the healing seen in the C57BL/6 ear hole by repair (*left*) versus the healing seen in the MRL ear hole by regeneration (*right*). In the C57BL/6 ear, a basement membrane is formed which separates the epidermis and dermis. In the MRL ear, the basement membrane forms but is broken down allowing epithelial–mesenchymal interactions and the proliferation of dermal fibroblasts and the growth of the blastema

key stage of the regenerative response is the ability for epithelial–mesenchymal interactions to occur, unimpeded by a barrier such as the basement membrane (Fig. 4). In the amphibian, the lack of a basement membrane in the healing wound (Repesh and Oberpriller 1980), whether never formed or continually being broken down, and the associated activity of proteases such as the matrix metalloproteinases (MMPs), has been shown to lead to the growth of the dermal cells below the epithelial cell layer and with it the formation of the blastema (Stocum and Dearlove 1972; Globus et al. 1980; Brockes 1997; Chernoff et al. 2000; Christensen et al. 2002), a group of undifferentiated cells that proliferates until the appropriate time when they stop and then re-differentiate into mature tissue (Grillo et al. 1968; Dresden and Gross 1970; Yang and Bryant, 1994; Miyazaki et al. 1996). Experiments have shown that premature induction of the basement membrane in the regenerating amphibian limb leads to the cessation of the regenerative process and the formation of scar tissue (Stocum and Crawford 1987).

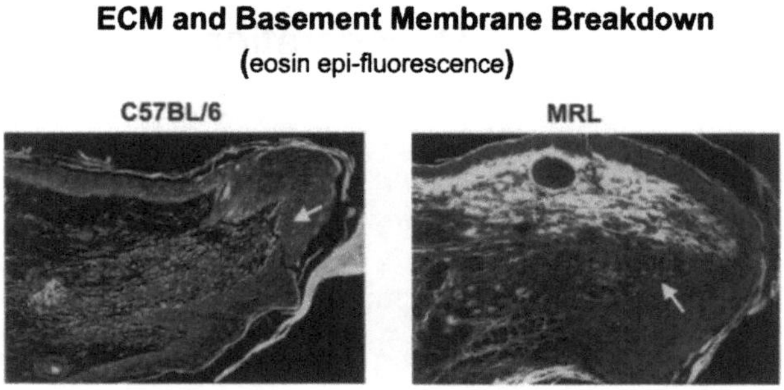

Fig. 5 Histological evidence of basement membrane breakdown. Ear punches were made in the center of the ear pinnae, creating 2-mm through-and-through wounds. Histological sections were made through the holes and stained with hematoxylin and eosin. Protein-bound eosin is seen by epifluorescence using a broad spectrum FITC filter. Sections made from ears injured 5 days previously can be seen from C57BL/6 (*left*) and MRL (*right*). The presence of the basement membrane between the epidermis and dermis can be clearly seen (*arrow*) in the C57BL/6 but not in the MRL (see Gourevitch et al. 2003)

In mammalian wound healing (repair), the process of wound coverage includes re-epithelialization that is slow compared to regenerating species. Next, a provisional matrix is laid down and this is followed by the formation of not only scar tissue but also a basement membrane between the epidermis and dermis (Clark 1996). Many of these changes are accompanied by remodeling and protease activity. Similar changes are seen in the C57BL/6 and in the MRL mouse ear hole wound. At approximately day 5 after injury, the MRL ear hole shows signs of blastema formation whereas the C57BL/6 does not (Fig. 2). The differences between these two cases were not visually obvious. However, using the ability of eosin to stain protein nonspecifically and to fluoresce, we showed that at approximately day 5 (Fig. 5) the basement membrane separating the epithelial layer from the fibroblast layer of cells in the C57BL/6 mouse was present and remained so, whereas the MRL basement membrane that was present 4 days after wounding and thus had reformed like the C57BL/6, was at day 5 difficult or impossible to visualize. A likely reason for the disappearance of the basement membrane was a remodeling response with an increase in protease activity. We examined this possibility and found that both mice expressed MMP-2 and MMP-9 (proteases that could breakdown the basement membrane at the wound site). The predominant MMP-positive cell types appeared to be cells related to the

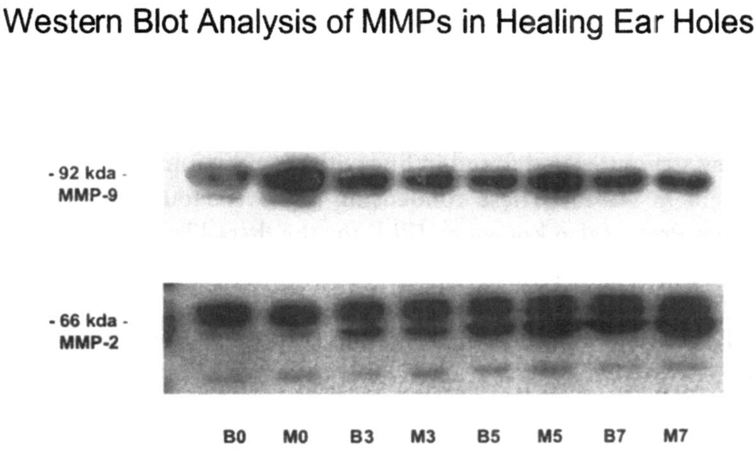

Fig. 6 The matrix metalloprotease response in the healing ear holes. Western blot analysis was carried out on denatured protein from healing tissue around the ear holes. Samples (12 μg total protein) were adjusted for protein concentration, run on SDS–PAGE, and blotted and stained with rabbit antibodies to either MMP-2 or MMP-9 (Sigma). Secondary horse radish peroxidase-labeled mouse anti-rabbit antibody (Sigma) was used and then developed with a chemiluminescence kit (Roche). Similar to the protease response seen using zymography (see Gourevitch et al. 2003), the protein levels were increased in the MRL as compared to the C57BL/6 healing ears, especially on days 5 and 7

inflammatory response and included mainly neutrophils and macrophages. Here, a difference was found between the MRL and the C57BL/6. First, there was more MMP-2 and MMP-9 expression in the MRL ear hole (Fig. 6). Second, the enzymatic activity was greater in the MRL, potentially due to the lower amount of TIMPs (tissue inhibitor of MMP) in the MRL ear hole (Gourevitch et al. 2003). Finally, more inflammatory cells overall were seen in the MRL ear after wounding. Thus, a likely explanation for basement membrane disappearance and concomitant blastema growth in the MRL was potent proteinase activity.

2.2 The Blastema

In the MRL ear hole, an ear hole blastema forms with the appearance of dividing cells that are presumably de-differentiated and potentially derived from fibroblasts, muscle, and cartilage. This circular bulge continues to enlarge and then elongate until the points touch and close the hole (Fig. 2). The maturation of cells in the blastema or re-differentiation is seen as a condensation of cells no longer part of the general population of blastema cells but rather a cell population that appears to be condensing into new structures and found in line with the remaining mature cartilage present in the ear. These condensing cells were found to express the molecule *Pref-1* (also known as *Dlk1*, for the delta-like 1 homolog in *Drosophila*) (Samulewicz et al. 2002), initially identified as a molecule containing EGF-like repeats and expressed by pre-adipocytes, a population of cells committed to becoming mature adipocytes but not yet fully differentiated (Smas and Sul 1993). Downregulation of *Pref-1* by such cells and other populations as well has been reported to be required for full differentiation (Garces et al. 1999) and this molecule also seems to be required for various stroma–cell interactions (Bauer et al. 1998). In the MRL ear hole, *Pref-1* RNA expression levels early after injury are low (but nevertheless greater than those seen in C57BL/6) but increase dramatically between days 12 and 14. The cells in the blastema eventually mature into new cartilage and muscle (Clark et al. 1998), though the molecular signals for such differentiation have not been identified (Fig. 3).

Pref-1 or Delta-like molecule 1 is similar to the molecule Delta, the ligand for Notch. Pref-1 could very well be part of the pathway beginning to be identified in the de-differentiating newt limb involving both radical fringe and Notch (Cadinouche et al. 1999; Kaneko et al. 2001).

3
A Model of Cardiac Regeneration

3.1
Cardiomyocyte Proliferation

The regeneration of mature myocardial tissue is generally not thought to occur in mammals. This cardiologic view is anchored by two principal assertions. First is the belief that adult mammalian cardiomyocytes are 'terminally' differentiated (Nadal-Ginard 1978; Scheuer and Buttrick 1987; Carbone et al. 1995), a position, which combines, perhaps inappropriately, physiological specialization (vis a vis the expression of a specific battery of genes) and an irreversible loss of proliferative capacity. However, adult cardiomyocyte proliferation has, in fact, been described in mammals (Polezhaev 1972; Oberpriller et al. 1985; Borisov

1998, 1999; Beltrami et al. 2001; Anversa and Nadal Ginard, 2002a). Secondly, a questionable comparison has been drawn to the regenerative mechanism observed for adult skeletal muscle. It has become popular to presume, to a greater rather than a lesser extent, that for striated muscle regeneration to proceed a reservoir of quiescent, stable, compartmentalized cells (called satellite cells) is activated in response to acute and chronic injury. The role of satellite cells in skeletal muscle regeneration is, at this point, obvious and thus beyond all reasonable contention (Mauro 1961; Moss and Leblond 1971; Zammit et al. 2002). The case for a putative cardiac stem cell population has not been clearly defined nor has it been convincingly established (Taylor et al. 2002, Anversa and Nadal Ginard 2002b). Moreover it has yet to be conclusively demonstrated that endogenous stem cells—whether resident in the heart or derived from a circulating population—are localized to an injury site or in fact participate in any injury response process.

It is, however, believed that adult cardiomyocyte proliferation does occur in newts along with cardiac injury repair and regeneration (Rumyantsev 1973; Oberpriller et al. 1985; Brockes and Kumar 2002). It has also been shown that striated muscle-derived newt myotubes display evidence of cell cycle progression in that their nuclei enter S phase in vitro after stimulation by high concentrations of serum (Lo et al. 1993; Echeverri et al. 2001; Odelberg et al. 2000). This is in contrast to mammalian C2C12 myotubes whose nuclei do not enter S phase under similar conditions (this is consistent with the in vivo results).

Given the ability of MRL to regrow tissue in a manner quite different from normal wound repair, we tested the ability of this mouse to heal myocardial wounds. An injury was made to the right ventricle (RV) of the heart using a cryoinjury model (Vracko and Thorning 1985; Taylor et al. 1998), an injury which is both controllable in size and leads to cardiomyocyte cell death (Fig. 7). Mice were given bromodeoxyuridine (BrdU), a nucleic acid analog which incorporates into DNA upon cell division (during S phase of the cell cycle), at the time of surgery and in the drinking water for 60 days.

Five days after such an injury, the MRL and C57BL/6 injury site showed a similar replacement of cardiac tissue with fibroblastic-like cells. With time, such tissue was replaced by scar tissue in the C57BL/6. In the MRL, however, such tissue began being replaced by cardiomyocytes and finally normal-looking tissue was seen with normal function (Leferovich et al. 2001). By day 60, the injury site in the MRL was filled

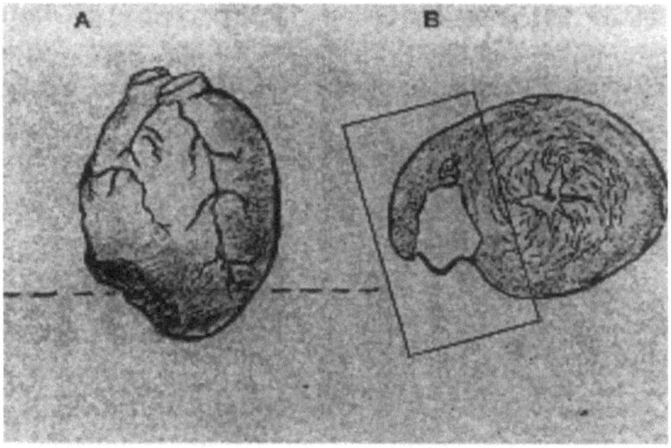

Fig. 7A, B Cryoinjury made to the right ventricle of the heart. A diagram of the injury made to the mouse heart shows the location of the injury (**A**) and a cross-section of the heart and the trans-mural area affected (**B**). A cryoinjury was made through the diaphragm to the right ventricle of the heart without opening the thoracic cavity. The *boxed area* in (**B**) is scarred in the C57BL/6 and filled with normal cardiomyocytes in the MRL. (From Leferovich et al. 2001)

with cardiomyocytes (Fig. 8) that were BrdU-positive, indicating cell division whereas the C57BL/6 had BrdU-positive fibrotic cells in the area of injury imbedded in the scar tissue. A mitotic index comparing the number of BrdU-positive cardiomyocyte nuclei/1,000 cardiomyocytes from the RV injury site showed 1%–2% for the C57BL/6 hearts and approximately 20% for the MRL hearts. By day 90, function in the MRL had returned to normal as determined by echocardiography. Animals examined by echocardiograms were followed for 1 year. The injured MRL tissue looked normal with little scar tissue as compared to the C57BL/6 tissue which was still completely scarred at the wound site. The level of cell proliferation in the MRL was similar to that seen in the amphibian (Borisov 1998) compared to that seen in the C57BL/6 (Leferovich et al. 2001), in other mammals (Polezhaev 1972; Borisov 1999) and in mammals including humans (Beltrami et al. 2001; Anversa and Nadal Ginard 2002a).

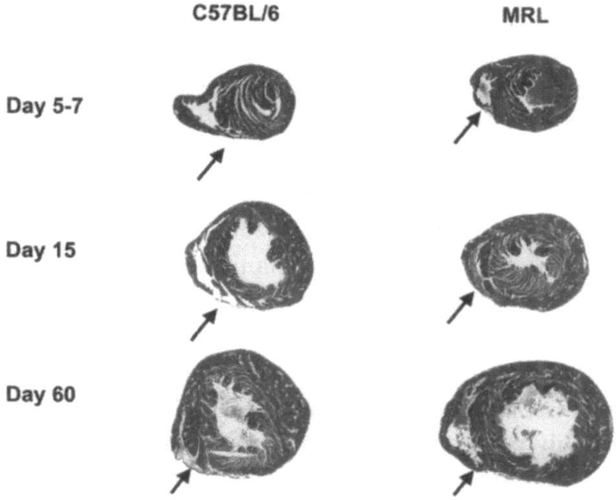

Fig. 8 The spontaneous regenerative response in the MRL mouse heart after injury. Light micrographs of trichrome-stained, transversely sectioned cryo-infarcted hearts 5 days, 15 days, and 60 days after cryoinjury with the nonhealer C57Bl/6 mouse hearts *(left)* and the healer MRL mouse hearts (*right*). The C57Bl/6 wound is predominantly scar tissue; the MRL wound myocardium appears normal with only a small amount of scar tissue on the epicardial surface. The site of the injury, the right ventricle, is indicated (*arrow*)

3.2
The Role of Stem Cells

One explanation for the presence of dividing cells in the heart is that there is a population of stem cells that can migrate to the heart, mature into cardiomyocytes, and divide. This has been shown experimentally with the transfer of bone marrow stem cells which contribute to the cardiomyocyte population in the mouse (Jackson et al. 2001; Orlic et al. 2001a,b; Toma et al. 2002). This is supported by data showing that in human transplant male patients, who have received female hearts, cardiomyocytes in varying degrees are Y chromosome-positive (Laflamme et al. 2002; Quaini et al. 2002). However, these findings have been challenged in studies showing that the Y chromosome-positive cells are inflammatory cells involved in allograft rejection (Taylor et al. 2002).

There is also evidence that there are resident cardiac stem cells (Warejcka et al. 1996; Anversa and Nadal-Ginard, 2002a).

Another intriguing possibility for cardiomyocyte replacement is the evidence showing that stem cells can recirculate and fuse with other cells and adopt their phenotype (Terada et al. 2002). And finally it has been shown that cardiomyocytes can induce the transdifferentiation of endothelial cells into cardiomyocytes (Condorelli et al. 2001).

Thus, there appear to be many possible mechanisms for cardiac replacement (Grounds et al. 2002). Unfortunately, cardiomyocyte repair in adults in mammals does not seem to occur spontaneously after injury as has been seen in the MRL mouse model. At present, there is no evidence of stem cell involvement in the MRL mouse but such studies are currently ongoing. There certainly is the view that epimorphic regeneration involves mainly de-differentiation of mature tissue and then redifferentiation back into mature tissue (Brockes and Kumar 2002) and may not involve stem cells.

In our studies on ear hole closure and with the possible role of epidermal stem cells contributing to this hole closure, preliminary studies have indicated that an in vitro epidermal stem cell model of R.J. Morris and colleagues (see the chapter by N.V. Popova and R.J. Morris, this volume; Morris and Potten 1994; Popova et al. 2002, 2003) shows that surprisingly MRL cultures have very few and very small epidermal stem cell colonies whereas C57BL/6 cultures show a robust response both in number and in size. How this will replicate what is seen in vivo remains to be determined.

3.3 Extracellular Matrix Breakdown

One reason for the lack of myocyte replacement after injury is the presence of scar tissue and the problem of matrix remodeling and protease activity in the heart (Dixon et al. 1997; Lee et al. 1998; Heymans et al. 1999; Li et al. 2000; Song et al. 2000; Sun and Weber 2002). Given the response to injury in the MRL ear and the lack of scarring, we examined the scar response in the MRL heart by determining the level of hydroxyproline as a measure of collagen protein deposition and scar tissue. In the C57BL/6 tissue, this level went up after injury. In the MRL, however, the amount of hydroxyproline continued to go down after injury, and remained low even at 60 days (Leferovich et al. 2001). Collagen type I levels as determined by reverse transcription–polymerase chain reaction showed that both strains of mice after injury upregulated collagen RNA synthesis. This suggested that a very likely explanation for the reduction in the collagen protein level was an increase in protease activity in the MRL heart after injury.

To determine how this occurs, we examined the MMP responses in these two animal models. MMP responses were again seen predominantly in an infiltrating population similar to that seen in the ear hole wounds (Gourevitch et al. 2003) with a response that was greater and of longer duration in the MRL injury (Bedelbaeva et al., manuscript in preparation).

3.4 The Angiogenic Response

It was proposed by Borisov (1999) that 'regenerative responses are limited by the capacity of the animal to resorb rapidly the foci of necrosis and to revascularize and reinnervate the volume of regenerating tissue'. It seems clear that the survival of cardiomyocytes following infarction requires a re-establishment of the vascular system (e.g., Sasayama and Fujita 1992; Battagay 1995; Kocher et al. 2001). Thus, the use of angiogenic factors to enhance cardiac healing after infarcts has been suggested (Folkman and Klagsbrun 1987; Isner 2002).

In the ear hole wound in the MRL mouse, we noticed that neovascularization or angiogenesis was quite rapid (Clark et al. 1998). Thus, we examined the heart for similar effects. Uninjured MRL and C57BL/6 heart had similar numbers of blood vessels. Upon injury, however, the angiogenic response was quite different with the level of angiogenesis coinciding with the level of the regenerative response seen (Bedelbaeva et al., unpublished results).

Interestingly, the remodeling process that occurs with scarring in the myocardial infarct affects angiogenesis (Nelissen-Vrancken et al. 1996; Kalkman et al. 1996) and has been shown to involve collagenase and other proteinases (Heymans et al. 1999), a process which appears to be quite different between MRL and C57BL/6 (discussed above).

4
The Role of Scar Tissue in Other Models of Regeneration

In recent studies carried out in our laboratory, a spinal cord injury involving complete transection of the mouse spinal cord in an environment of limited scarring resulted in quite significant recovery of function (Seitz et al. 2002). This study reveals the extensive ability of the mammal to regenerate in the absence of scar tissue. It also suggests that the property of the MRL mouse to have enhanced ability to breakdown scar tissue or not allow it to form (Clark et al. 1998; Gourevitch et al. 2003) may be directly related to its ability to regenerate. In an amphibian regenerating spinal cord injury model, proteases play an important part in the regenerative process (Chernoff et al. 2000, 2002). Recent data from many laboratories have supported the idea that scar removal allows a regenerative response (Davies et al. 1999; Stichel et al. 1999; Bradbury et al. 2002; Condic and Lemons 2002).

A molecular examination of healing spinal cord injury showed that *apolipoprotein E* (*apo-E*) a molecule which has been shown to be involved in lipid metabolism, the immune response and inflammation, and diseases of the central nervous system such as Alzheimer's and multiple sclerosis, is upregulated in the spinal cord healing and peaks at day 10 after injury. Besides expression in neurons and astrocytes, *apo-E* appears in neutrophils and macrophages early after injury (Seitz et al. 2003). It is not clear if this molecule is involved with scar formation and this is currently being explored.

5
The Genetics of Regeneration

As discussed above, the examples of regeneration in mammals is quite limited. In the ear hole model of regeneration, there is the rabbit model and the MRL mouse model. Unlike the rabbit, however, the mouse provides a perfect model to identify genes given the sequencing of the genome, the availability of inbred strains, the ease of breeding of mouse lines, and the mapping tools for identification of loci, genes and gene products. Studies on the genetics of the ear hole closure response has indicated the heritability of this phenotype and in a genomewide screen using microsatellite mapping using over 300 (MRLxC57BL/6)F2 and BC mice, multiple loci on chromosomes 7, 8, 12, 13 and 15 were identified

(McBrearty et al. 1998; Blankenhorn et al. 2003). Furthermore, the generation of healer phenotype congenic mice have clearly shown that this trait is indeed multigenic (E. Heber-Katz, unpublished results). The mapping also revealed that the *fas* gene is not involved in the healing phenomenon since the same degree of ear hole closure was seen in the MRL $lpr^{+/+}$ (MRL/MpJ) strain.

We have recently studied the effect of sex on healing—the sexual dimorphism of this trait. We found that female mice in general heal better than male mice, and mice differ by sex with respect to which healing genes are important in their wound closure (Blankenhorn et al. 2003). Furthermore, castration experiments led to more complete ear hole closure in male mice, in which they achieved the level of female healing at the end of 30 days. It has also been shown that wound healing is controlled by different genes in male and female mice using the F2 crosses described above.

A search for further loci using mouse strains crossed to the MRL mouse and examining the ear hole closure phenotype revealed, in fact, more activity at other chromosomal locations. This was true for F2 mice generated from MRL and *M.m.castaneus* matings (E. Heber-Katz et al. unpublished results) as well as for F2 mice generated from MRL×SJL matings (Masinde et al. 2001). From these studies, two sets of nomenclature have appeared. One for the MRL×C57BL/6 and MRL×CAST/EI crosses and called *heal* loci (McBrearty et al. 1998; Blankenhorn et al. 2003; E. Heber-Katz et al. unpublished results) and one for the MRLxSJL crosses and called *Sth* or soft tissue healing loci (Masinde et al. 2001). The current loci which are 14 or greater in number at the present time and can be seen in Table 1 will also be available under 'heal' at http://www.informatics.jax.org/searches/marker_form.shtml.

6
Conclusion

There is an implicit assumption in biology that there is an accelerating loss of plasticity in concert with evolution due to selective pressures. Thus, earlier forms often show multiple pathways to accomplish some very basic life functions. For example, certain forms of algae such as *Chlamydomonas rheinhardtii* can use either photosynthesis or chemical oxidation, depending on the availability of light. The slime mold *Dictyostelium discoideum* can exist as a single celled amoeba, or form a multi-

Table 1 Heal quantitative trait loci in all strains. *Heal* designations are as seen in McBrearty et al. 1998, Blankenhorn et al. 2003, and E. Heber-Katz et al. (unpublished results). *Sth* designations are as seen in Masinde et al. 2001

Peak Marker	cM	Name of healing QTL	Origin of healer in MRL×B6	Origin of healer in MRL×CAST
D8Mit211	49	*Heal1*	B6 M>F	
D13Mit15, 16	9–13	*Heal2*	MRL, M>F	MRL, M>F
D13Mit228	60–65	*Heal3/Sth10*	MRL	
D15Mit244	57	*Heal4*	MRL	
D12Mit136	13	*Heal5*	B6	
D7Mit85	27–52	*Heal6*	MRL	
D13Mit245–D13Mit139	30–36	*Heal7*	MRL, M>F	MRL, M>F
D4Mit148	66–71	*Heal8*	MRL	
D18Mit123	30–44	*Heal9*	B6, F>M	
D11Mit213-61	58–70	*Heal10*	MRL, males	MRL, M>F
D16Mit122–110	4–20	*Heal11*	het	
D14Mit233	20	*Heal12*		CAST/Ei - in males
D17Mit93	44	*Heal13*		CAST/Ei
D1Mit334	49.7	*Sth1*		
D3Mit217	nd	*Sth2*		
D4Mit214	18	*Sth3*		
D4Mit31	51	*Sth4*	MRL	MRL
D6Mit261	37	*Sth5*		
D7Mit220	52	*Sth6*	BC1 suggestive	
D7Mit12	66	*Sth7*		
D9Mit207	33	*Sth8*		
D9Mit270	43	*Sth9*		Het, in females
D2Mit354	0–5	Suggestive		Het

cellular form called a slug, alternating sexual and vegetative reproductive cycles according to environmental conditions. Bakers yeast, *Saccharomyces cerevisae*, and the nematode worm, *Caenorhabditis elegans*, are capable of both sexual and asexual reproduction and there are some fish species, such as the Hawaiian protogynous Wrasse (*Thalassoma duperrey*) with individuals that can switch from male to female forms depending on mating pressures. Also, the parthogenetic Southwest Whiptail lizard

Cnemidophorus uniparens is a species of female clones that adopt male or female sexual behavior according to environmental conditions. To a great extent, these properties have been lost in mammals (although note: metabolic adaptation in tissues and organ systems still occurs according to environmental conditions and states of cellular injury, as tissues shift between anaerobic and aerobic energy production in different physiological states).

The phenomenon of decreasing plasticity certainly characterizes observations in wound healing, with multi-organ regeneration typical in lower forms lost to scarring in higher forms. In mammals, it appears that scarring has all but totally supplanted epimorphic regeneration. And yet, there are surprising counter examples, i.e., antlers and complete closure of rabbit ear holes. However, that a mouse strain is capable of wound healing in many ways as profound as the axolotl or newt is most surprising and calls into question the whole notion of an 'evolutionary arrow' pointing away from regeneration. Spallanzani would have liked this.

Acknowledgements. The work described here was generously supported by grants from the National Institutes of Health, the G. Harold and Leila Y. Mathers Foundation, and the F.M. Kirby Foundation. We would like to thank David Stocum and Robert Naviaux for useful discussions and to Elizabeth P. Blankenhorn and David Emanuel Sarfatti for their support and input.

References

Anversa P, Nadal-Ginard B (2002) Myocyte renewal and ventricular remodelling. Nature 415:240–243

Anversa P, Nadal-Ginard B. (2002a) Cardiac chimerism: methods matter. Circulation. 106:129–31

Adachi M, Watanabe-Fukunaga R, Nagata S. (1993) Aberrant transcription caused by the insertion of an endogenous retrovirus in an apoptosis gene. Proc. Natl. Acad. Sci. USA 90:1756–1760

Battegay EJ. (1995). Angiogenesis: mechanistic insights, neovascular diseases, and therapeutic prospects. J Mol Med. 73:333–46

Bauer SR, Ruiz-Hidalgo MJ, Rudikoff EK, Goldstein J, Laborda J. (1998) Modulated expression of the epidermal growth factor-like homeotic protein dlk influences stromal-cell-pre-B-cell interactions, stromal cell adipogenesis, and pre-B-cell interleukin-7 requirements. Mol Cell Biol 1998; 18:5247–55

Beltrami, A.P., Urbanek, K., Kajstura, J., Yan,S-M, Finato, N., Bussani, R., Nadal-Ginard, B., Silvestri, F., Leri, A, Beltrami, C.A., and Anversa, P. (2001) Evidence that human cardiac myocytes divide after myocardial infarction. New Engl J. Med. 344:1750–1757

Blankenhorn, E.P., Troutman, S., Desquennes-Clark, L., Zhang, X-M., and Heber-Katz, E. (2003) Sexually dimorphic genes regulate healing and regeneration in the MRL/MpJ mouse. Mammalian Genome. 14:250–260

Borgens, R.B. (1982) Mice regrow the tips of the foretoes. Science 217:747–50

Borisov, A.B. (1998) Cellular mechanisms of myocardial regeneration in *Cellular and Molecular Basis of Regeneration from Invertebrates to Humans.* eds. Ferretti, P and Geraudie, J. (John Wiley and Sons Ltd., Chichester,). pp.335–354

Borisov, A. (1999) Regeneration of skeletal and cardiac muscle in mammals: do non-primate models resemble human pathology? Wound Rep. Regen.7: 26–35

Bradbury EJ, Moon LD, Popat RJ, King VR, Bennett GS, Patel PN, Fawcett JW, McMahon SB (2002) Chondroitinase ABC promotes functional recovery after spinal cord injury. Nature 416:636 40

Brockes J.P. (1997) Amphibian limb regeneration: rebuilding a complex structure. Science 276:81–87

Brockes JP, Kumar A. (2002) Plasticity and reprogramming of differentiated cells in amphibian regeneration.Nat Rev Mol Cell Biol. 3:566–74

Cadinouche MZ, Liversage RA, Muller W, Tsilfidis C.(1999) Molecular cloning of the Notophthalmus viridescens radical fringe cDNA and characterization of its expression during forelimb development and adult forelimb regeneration. Dev Dyn. 214:259–68

Carbone, A, Minieri, M, Sampaolesi, M, Fiaccavento, R., De Feo, A., Cesaroni, P, Peruzzi, G., Di Nardo, P. (1995). Hamster Cardiomyocytes: a model of myocardial regeneration? Annals of the New York Academy of Sciences. 752:75–71

Chernoff EAG, O'Hara CM, Bauerle B, Bowling M. (2000) Matrix metalloproteinase production in regenerating axolotl spinal cord. Wound Repair Regen 8:282–291

Chernoff E, Sato K, Corn A, Karcavich R. (2002) Spinal cord regeneration: intrinsic properties and emerging mechanisms. Semin Cell Dev Biol. 13:361

Christensen RN, Weinstein M, Tassava RA. (2002) Expression of fibroblast growth factors 4, 8, and 10 in limbs, flanks, and blastemas of Ambystoma. Dev Dyn. 223:193–203

Clark, L., Clark, R., and Heber-Katz, E. (1998) A new model for mammalian wound repair and regeneration. Clin. Immun. Immunopath. 88:35–45

Clark, R.A.F. (1996) Wound repair: overview and general consideration. In: The Molecular and Cellular Biology of Wound Repair. (R.Clark, ed.). Plenum Press, NY. pp.3–35

Cohen, P.L. and Eisenberg, R.A. (1991). Lpr and gld: Single Gene models of Systemic Autoimmunity and Lymphoproliferartive disease. Annu. Rev. Immunol. 9,243–69

Condic ML, Lemons ML (2002) Extracellular matrix in spinal cord regeneration: getting beyond attraction and inhibition. Neuroreport 13:A37–48

Condorelli G, Borello U, De Angelis L, Latronico M, Sirabella D, Coletta M, Galli R, Balconi G, Follenzi A, Frati G, Cusella De Angelis MG, Gioglio L, Amuchastegui S, Adorini L, Naldini L, Vescovi A, Dejana E, Cossu G (2001) Cardiomyocytes induce endothelial cells to trans-differentiate into cardiac muscle: implications for myocardium regeneration. Proc Natl Acad Sci U S A 98:10733–8

Davies SJ, Goucher DR, Doller C, Silver J (1999) Robust regeneration of adult sensory axons in degenerating white matter of the adult rat spinal cord. J Neurosci 19:5810–5822

Dickie MM. Keeping Records, p25. In Biology of the Laboratory Mouse. Green EL (ed), Dover Publications. New York, 1975

Dinsmore, CE. (ed) (1991) A History of Regeneration Research: Milestones in the Evolution of a Science. Cambridge University Press. New York

Dixon, IMC, Haisong, J, Reid, NL, Scammell-La Fleur, T, Werner, JP, and Jasmin, G. (1997) Cardiac collagen remodeling in the cardiomyopathic syrian hamster and the effect of losartan. J. Mol Cell Cardiol. 29: 1837–1850

Douglas BS (1972) Conservative management of guillotine amputations of the fingers of children. Austral Paediatr J 8:86–90

Dresden MH, Gross J. (1970) The collagenolytic enzyme of the regenerating limb of the Newt Triturus viridescens. Dev Biol. 22:129–37

Echeverri K, Clarke JDW, and Tanaka E. (2001) In vivo Imaging indicates muscle fiber dedifferentiation is a major contributor to the regenerating tail blastema. Developmental Biology 236:151–164

Folkman, J. and Klagsbrun, M. (1987) Angiogenic factors. Science 235, 442–447

Garces C, Ruiz-Hidalgo MJ, Bonvini E, Goldstein J, Laborda J. (1999) Adipocyte differentiation is modulated by secreted delta-like (dlk) variants and requires the expression of membrane-associated dlk. Differentiation 64:103–14

Gardiner, D.M. and Bryant, S.V. (1996) Molecular mechanisms in the control of limb regeneration: the role of homeobox genes. Int. J. Dev. Biol. 40:797–805

Globus M, Vethamany-Globus S, Lee YCI. (1980) Effect of apical epidermal cap on mitotic cycle and cartilage differentiation in regeneration blastemata in the newt. Notophthalmus viridescens. Develop Biol 75:358–372

Gourevitch,D, Clark,L, Chen P, Seitz A, Samulewicz S, and E. Heber-Katz. (2003) Matrix Metalloproteinase Activity Correlates with Blastema Formation in the Regenerating MRL Mouse Ear Hole Model. Dev Dyn. 226:377–287

Grillo HC, Lapiere CM, Dresden MH, Gross J. (1968) Collagenolytic activity in regenerating forelimbs of the adult newt. Dev Biol 17:571–583

Gross, J. (1996) Getting to mammalian wound repair and amphibian limb regeneration: a mechanistic link in the early events. Wound Repair Reg. 4:190–202

Grounds MD, White JD, Rosenthal N, Bogoyevitch MA (2002) The role of stem cells in skeletal and cardiac muscle repair. J Histochem Cytochem. 50:589–610

Goss, R.J. (1970) Problems of antlerogenesis. Clin. Orthopaedics 69:227–238

Goss, R.J. and Grimes, L.N. (1975) Epidermal downgrowths in regenerating rabbit ear holes. J.Morphol. 146:533–542

Gourevich, D, Clark, L, Chen P, Seitz A, Samulewicz S, and E. Heber-Katz. (2003) Matrix Metalloproteinase Activity Correlates with Blastema Formation in the Regenerating MRL Ear Hole Model. Developmental Dynamics. 226: 377–387

Heymans S, Luttun A, Nuyens D, Theilmeier G, Creemers E, Moons L, Dyspersin GD, Cleutjens JP, Shipley M, Angellilo A, Levi M, Nube O, Baker A, Keshet E, Lupu F, Herbert JM, Smits JF, Shapiro SD, Baes M, Borgers M, Collen D, Daemen MJ, Carmeliet P. (1999) Inhibition of plasminogen activators or matrix metalloproteinases prevents cardiac rupture but impairs therapeutic angiogenesis and causes cardiac failure. Nature Medicine 5:1135–1142

Heber-Katz, E. (1999) The regenerating mouse ear. Sem. Cell Dev. Biol. 10:415–420

Heber-Katz, E., Chen, P., Clark, L., Zhang, X-M., Troutman, S., and Blankenhorn, EP. (2002) Novel genetic loci associated with wound healing and regeneration that distinguish MRL and M. m. castaneus mice. Submitted

Illingworth CM (1974) Trapped fingers and amputated finger tips in children. J Pediatr Surg 9:853–58

Isner, JM. (2002) Myocardial gene therapy. Nature 415:234–239

Jackson KA, Majika SM, Wang H, Pocius J, Hartley CJ, Majesky MW, Entman ML, Michael LH, Hirschi KK, Goodell M (2001) Regeneration of ischemic cardiac muscle and vascular endothelium by adult stem cells. J Clin Inv 107:1395–1402

Kalkman EA, Bilgin YM, van Haren P, van Suylen RJ, Saxena PR, Schoemaker RG. (1996) Determinants of coronary reserve in rats subjected to coronary artery ligation or aortic banding. Cardiovasc.Res. 32, 1088–1095

Kaneko Y, Hirota K, Matsumoto G, Hanyu Y. (2001) Expression pattern of a newt Notch homologue in regenerating newt retina. Brain Res Dev Brain Res. 128:53–62

Kench, J.A., Russell, D.M., Fadok, V.A., Young, S.K., Worthen, G.S., Jones-Carsen, J., Henson, J.E., Henson, P.M., and Nemazee, D. (1999) Aberrant wound healing and TGF-beta production in the autoimmune-prone MRL/+ mouse. Clin. Immunol. 92:300–310

Kocher AA, Schuster MD, Szabolcs MJ, Takuma S, Burkhoff D, Wang J, Homma S, Edwards NM, Itescu S. (2001) Neovascularization of ischemic myocardium by human bone marrow-derived angioblasts prevents cardiomyocyte apoptosis, reduces remodeling, and improves cardiac function. Nature Medicine 7:430–436

Laflamme MA, Myerson D, Saffitz JE, Murry CE. (2002) Evidence for cardiomyocyte repopulation by extracardiac progenitors in transplanted human hearts. Circ Res 90:634–40

Lee, JK, Zaidi, SHE, Liu, P., Dawood, F, Cheah, AYL, Wen, W-H, Saiki, Y., and Rabinovitch, M. (1998) A serine esterase inhibitor reduces inflammation and fibrosis and preserves cardiac function after experimentally-induced murine myocarditis. Nature Medicine 4: 1383–1391

Leferovich, J., Bedelbaeva, K., Samulewicz, S, Xhang, X-M, Zwas, DR, Lankford, EB, and Heber-Katz, E. (2001) Heart regeneration in adult MRL mice. Proc. Natl. Acad. Sci. USA, 98:9830–9835

Li, X., Gu, W., Masinde, G., Hamilton-Ulland, M., Xu, S., Mohan, S., and Baylink, D.J. (2001) Genetic control of the rate of wound healing in mice. Heredity. 86:668–674

Li, YY, Feng,YQ, Kadokami,T, McTiernan, CF, Draviam,R, Watkins, SC, and Feldman AM. (2000) Mycardial extracellular matrix remodeling in transgenic mice overexpressing tumor necrosis factor alpha can be modulated by anti-tumor necrosis factor alpha therapy. PNAS 97:12746–12751

Lo D, Allen F, Brockes JP. (1993) Reversal of muscle differentiation during urodele limb regeneration. PNAS 90:7230–7234

Mauro, A. 1961. Satellite cell of skeletal muscle fibres. J Biophys Biochem Cytol 9:493 498

Masinde, G.L., Li, X., Gu, W., Davidson, H., Mohan, S., and Baylink, D.J. (2001) Identification of wound healing/regeneration quantitative trait loci (QTL) at multiple time points that explain seventy percent of variance in (MRL/MpJ and SJL/J) mice F2 population. Genome Res. 11:2027–2033

McBrearty, B.A., Desquenne-Clark, L., Zhang, X-M., Blankenhorn, E.P., and Heber-Katz, E. (1998) Genetic analysis of a mammalian wound healing trait. Proc. Natl. Acad. Sci. USA, 95:11792 - 11797

Miyazaki, K,Uchiyawa K, Imokawa Y, Yoshizato K. (1996) Cloning and characterization of cDNAs for matrix metalloproteinases of regenerating newt limbs. Proc Natl Acad Sci USA 93:6819–6824

Morris RJ, and Potten CS. (1994) Slowly cycling label-retaining epidermal cells behave like clonogenic stem cells in vitro Cell Prolif. 27:279–289

Moss and Leblond (1971) Satellite cells as the source of nuclei in muscles of growing rats. Anat Rec. 170:421–35

Murphy, E.D. and Roths, J.B. (1979) Autoimmunity and lymphoproliferation: Induction by mutant gene lpr and acceleration by a male-associated factor in strain BXSB. in*Genetic Control of Autoimmune Disease.* (eds. Rose, N.R., Bigazzi, P.E., and Warner, N.L.) 207–220 (Elsevier, New York)

Nadal-Ginard,B. (1978) Comittment, fusion, and biochemical differentiation of a myogenic cell line in the absence of DNA synthesis. Cell. 15:855–864

Nelissen-Vrancken HJ, Debets JJ, Snoeckx LH, Daemen MJ, Smits JF. (1996) Time-related normalization of maximal coronary flow in isolated perfused hearts of rats with myocardial infarction. Circulation. 1996 93:349–55

Neufield DA (1989) Epidermis, basement membrane, and connective-tissue healing after amputation of mouse digits: implications for mammalian appendage regeneration. Anat Rec 223:425–32

Newmark PA, Sanchez Alvarado A. (2002) Not your father's planarian: a classic model enters the era of functional genomics. Nat Rev Genet. 3:210–9

Oberpriller,JO, Ferrans, VJ, McDonnell, TJ, and Oberpriller,JC. (1985) Activation of DNA synthesis and mitotic events in atrial myocytes following atrial and ventricular injury. in *Pathobiology of Cardiovascular Injury.* Ed.Stone and Weglicki, (Martinus Nijhoff Publishing, Boston) pp. 410–421

Odelberg SJ, Kollhoff A, and Keating MT. (2000) Dedifferentiation of Mammalian Myotubes Induced by msx1. Cell 103:1099–1109

Orlic, D, et al. (2001a) Bone marrow cells regenerate infarcted myocardium. Nature.410:701–5

Orlic D et al. (2001b) Mobilized bone marrow cells repair the infarcted heart, improving function and survival. PNAS. 98:10344–10349

Polezhaev, L.V. (1972) Loss and Restoration of Regenerative Capacity in *Tissues and Organs of Animals* ed. Polezhaev LV (Harvard University Press, Cambridge).pp. 153–199

Popova, NV, Tryson K, Wu K, and Morris RJ. (2002) Evidence that keratinocyte colony number is genetically controlled. Exp Dermatol. 11:503–508

Popova NV, Teti KA, Wu KQ, Morris RJ.Identification of two keratinocyte stem cell regulatory loci implicated in skin carcinogenesis.Carcinogenesis (2003) 24(3):417–25

Quaini F, Urbanek K, Beltrami AP, Finato N, Beltrami CA, Nadal-Ginard B, Kajstura J, Leri A, Anversa P. (2002) Chimerism of the transplanted heart. N Engl J Med. 346:5–15

Quinones JL, Rosa R, Ruiz DL, Garcia-Arraras JE. (2002) Extracellular matrix remodeling and metalloproteinase involvement during intestine regeneration in the sea cucumber Holothuria glaberrima. Dev Biol. 250:181–97

Reginelli AD, Wang YQ, Sassoon D, Muneoka K.1995. Digit tip regeneration correlates with regions of Msxl (Hox 7) expression in fetal and newborn mice regrowth. Development 121:1065–76

Repesh LA, Oberpriller JC. (1980) Ultrastructural studies on migrating epidermal cells during the wound healing stage of regeneration in the adult newt, Notophthalmus viridescens. Am J Anat. 159:187–208

Rumyantev, PP. (l973) Post-injury DNA synthesis, mitosis, and ultrastructural reorganization of adult frog cardiac myocytes. Z. Zellforsch. 139: 431–450

Samulewicz, SJ, Clark,L, Seitz,A., and E. Heber-Katz. (2002) Expression of Pref-1, A Delta-Like Protein, in Healing Mouse Ears. Wound Repair and Regeneration, 10:215–221

Sasayama, S. and Fujita, M. (1992) Recent insights into coronary collateral circulation. Circulation, 85:1197–1204

Sawada T, Oofusa K, Yoshizato K. (1999) Characterization of a collagenolytic enzyme released from wounded planarians dugesia japonica. Wound Rep Regen 7:458–466

Scheuer J, Buttrick P. (1987) The cardiac hypertrophic responses to pathologic and physiologic loads. Circulation. 75:I63–8

Seitz, A.,Aglow, E., and Heber-Katz, E. (2002) Recovery from spinal cord injury: A new transection model in the C57Bl/6 mouse. J. Neuroscience Research. 67:337–345

Seitz,A, Kragol,M, Aglow,E, Showe, L. and E, Heber-Katz. (2003) Apo-E expression after spinal cord injury in the mouse. J. Neuroscience Research. 71:417–426

Smas CM, Sul HS. (1993) Pref-1, a protein containing EGF-like repeats, inhibits adipocyte differentiation. Cell 73:725–34

Stichel CC, Niermann H, D'Urso D, Lausberg F, Hermanns S, Muller HW (1999) Basal membrane-depleted scar in lesioned CNS: characteristics and relationships with regenerating axons. Neurosci 93:321–33

Song, W.,Jackson,K., and McGuire, P.G. (2000) Degradation of Type IV Collagen by Matrix Metalloproteinases is an important step in the epithelial-mesenchymal transformation of the endocardial cushions. Develop Biol. 227:606–617

Stocum, D.L. (1984) The urodele limb regeneration blastema. Determination and organization of the morphogenetic field. Differentiation 27:13–28

Stocum, D.L. (1996) Tissue restoration: approaches and prospects. Wound Repair Reg. 4:3-15

Stocum DL and Crawford, K (1987) Use of retinoids to analyze the cellular basis of positional memory in regenerating amphibian limbs. Biochem Cell Biol. 65:750–61

Stocum DL and Dearlove GE. (1972) Epidermal-mesodermal interaction during morphogenesis of the limb regeneration blastema in larval salamanders. J. Exp. Zool. 181:49–62

Sun, Y, and Weber, KT. (2000) Infarct scar: a dynamic tissue. Cardiovacular Research. 46:250–6

Tabin, C.J. (1991) Retinoids, homeoboxes, and growth factors: towards molecular models for limb development. Cell 66:199–217

Taylor, D.A., Atkins, B.Z., Hungspreugs, P., Jones, T.R., Reedy, M.C., Hutcheson, K. A., Glower, D.D. and Kraus, W.E. (1998) Regenerating functional myocardium: improved performance after skeletal myoblast transplantation. Nature Medicine, 4: 929–933

Taylor DA, Hruban R, Rodriguez ER, Goldschmidt-Clermont PJ (2002) Cardiac chimerism as a mechanism for self-repair: does it happen and if so to what degree? Circulation. 106:2–4

Technau U, Cramer von Laue C, Rentzsch F, Luft S, Hobmayer B, Bode HR, Holstein TW (2000) Parameters of self-organization in Hydra aggregates. Proc Natl Acad Sci 97:12127–31

Terada N, Hamazaki T, Oka M, Hoki M, Mastalerz DM, Nakano Y, Meyer EM, Morel L, Petersen BE, Scott EW (2002) Bone marrow cells adopt the phenotype of other cells by spontaneous cell fusion. Nature 416:542–5

Theofilopoulos, A.N. Immunologic genes in mouse lupus models (1993) in:*The Molecular Pathology of Autoimmune Diseases.* (eds. Bona, C., Siminovitch, K.A., Zanetti, M. and Theofilopoulos, A.N.) 281–316 (Harwood Academic Publishers, Langhorne)

Toma,C, Pittenger, MF, Cahill KS, Byrne BJ, Kessler P (2002) Human mesenchymal stem cells differentiate to a cardiomyocyte phenotype in the adult murine heart. Circulation. 105:93–98

Vracko, R. and Thorning, D (1985) Freeze-thaw injury of rat heart across an intact diaphragm: a new model for the study of the response of myocardium to injury. Cardiovasc Res., 19:76–84

Watanabe-Fukunaga, R., Brannan, C., Copeland, N.G., Jenkins, N.A., and Nagata, S (1992) Lymphoproliferation disorder in mice is explained by defects in Fas antigen that mediates apoptosis. Nature 356:314–316

Watson ML, Rao JK, Gilkeson GS, Ruiz P, Eicher EM, Pisetsky DS, Matsuzawa A, Rochelle JM, Seldin MF (1992) Genetic analysis of MRL-lpr mice: relationship of the Fas apoptosis gene to disease manifestations and renal disease modifying loci. J. Exp. Med. 176, 1645–1656

Warejcka DJ. Harvey R, Taylor BJ. Young HE, Lucas PA (1996) A population of cells isolated from rat heart capable of differentiating into several mesodermal phenotypes. J Surg Res 62:233–242

Yang EV, Bryant SV (1994) Developmental regulation of a matrix metalloproteinase during regeneration of axolotl appendages. Dev Biol 166:696–703

Zammit PS, Heslop L, Hudon V, Rosenblatt JD, Tajbakhsh S, Buckingham ME, Beauchamp JR, Partridge TA (2002) Kinetics of myoblast proliferation show that resident satellite cells are competent to fully regenerate skeletal muscle fibers. Exp Cell Res 281:39–49

Subject Index

A
acid hydrolase 9–10
adult stem cell 104
AEC factor 14
affinophoresis assay 35, 40
amphibian 71–73, 84, 165, 170–171, 176, 180
aneurogenic limb 23–24
angiogenesis 179
anteroposterior axis 29, 43–44
antler 166, 183
apical ectodermal ridge 5
apical wound epidermis 5
apoptosis 19, 26

B
banded hedgehog 44
basement membrane 170, 172
blastema 168, 170–173
BrdU 175–176

C
C2C12 mouse myotubes 11–12, 51
C57BL/6 mouse 172
cardiac myocyte / cardiomyocyte 140, 142–144, 150–151, 153, 155, 166, 174–175, 177–179
cartilage 165, 169, 173
chimeric receptor construct 38
collagen type I 178
compensatory hyperplasia 3–4
congenital anomaly 149
cytoplasmic retinoic acid binding protein 38

D
dedifferentiation 2, 4–5, 8–10, 12–13, 15, 17–19, 27, 30, 32, 46, 48–49, 51–52
degradation of the ECM 9, 11
Delta 174
denervation 19–22
developmental plasticity 25, 52
differential affinity 35
digit 166
distal boundary 31–33
disulfram 48
dorsal iris cell 48–49
dorsoventral axis 44

E
En-1 41, 45
Ephrin A receptor 37
epiblastic-like 91
events of limb regeneration 5

F
fibroblast growth factor 2, 16
fibroblast growth factor-2 10
Fibronectin 13, 17, 23

G
glial growth factor 2 21

H
histolysis 2, 8, 15, 19, 22, 52
HoxA-D gene 41
human 72, 85, 87–89, 91, 97, 104
hyaluronate 13, 20, 23
hydroxyproline 178

I
induction of regeneration 50–51, 53
inhibitor of differentiation gene 19
insulin-like growth factor-I 10
intercalary regeneration 29–31, 34, 36, 39, 42
intestine 4, 45

J

jaw 4, 45

K

keratinocyte stem cell 111–114, 123, 128–129, 131, 133–134

L

lens 2, 4–5, 46–48

LG/J mouse 170

limb regeneration 5, 7–8, 11, 23, 26–28, 43–46

Lmx-1 41, 44–45

M

macrophage 166, 173, 180

mammal 85

matrix metalloproteinase (MMP) 9, 171, 179

mechanism of Blastema Formation 8

mesenchymal 71, 85, 87–89, 93, 97, 104

microsatellite mapping 180

MRL mouse 165, 169, 172, 177–181

msxl gene 12

N

neovascularization 169, 179

nerve 2–6, 8, 12, 14–15, 17, 19–25, 41, 46, 52

nerve mitogenic factor 20

neural factor 19

neural retina 2, 4–5, 46–47, 51

neurotrophic factor 22

neutrophil 166, 173, 180

Notch 174

Notch pathway 13

NvTboxl 43

P

Pax-6 48

Pelizaeus–Merzbacher disease 49

personal ESC line 49

pigmented epithelial cell 2, 46

plasticity of Blastema Cell 26

platelet-derived growth factor 10

pluripotent 71, 73, 84–85, 87–89, 91, 93, 97, 101–104

positional identity 2, 30, 34–35, 37–38, 40–41, 43, 52

Pref-1 174

progenitor Cell 143, 149–150

protease 171–172, 178, 180

proximodistal Axis 41

R

RAR $\delta 2$ 38, 40

RAR isoform 37

re-epithelialization 170, 172

regeneration 140, 153–156, 158

regenerative biology and medicine 4, 51

regenerative medicine 2, 4, 49–50

repair 170–172, 175, 178

reserve stem cell 3–4, 25, 27–28, 50

retinoic acid 10, 34, 39

retinoic acid receptor 37, 39

retinoic acid response element 37

retinoid X receptor 37

rfrng 12–13, 40

rise to teratocarcinoma 49

rule of least (shortest) intercalation 31

S

scar 170, 172, 175, 178, 180

self-organization 30, 32–33, 42, 45, 52

sexual dimorphism 181

skin carcinogenesis 112, 114, 131, 133–134

sonic hedgehog 16, 43

Spallanzani 165–167, 169, 171, 173, 175, 177, 179, 181, 183, 185, 187, 189

spatial Organization of Tissue Pattern 28

spinal cord injury 180

stem cell 140, 143, 147, 153, 175, 177–178

substance P 2, 12, 21, 25, 52

T

thrombin 11, 52

tissue engineering 140, 146–150, 152, 158

tissue inhibitor of metallo-proteinase (TIMP) 10

transferrin 2, 12, 14, 21, 25, 52
transforming growth factor β 10
transplantation 139–147, 152, 155, 158
tumor necrosis factor α 10

W
Wnt-7 41
wound epithelium 2, 16

X
Xenopus laevis 7, 46

Z
ZPA 40, 43

Current Topics in Microbiology and Immunology

Volumes published since 1989 (and still available)

Vol. 236: **Kraehenbuhl, Jean-Pierre; Neutra, Marian R. (Eds.):** Defense of Mucosal Surfaces: Pathogenesis, Immunity and Vaccines. 1999. 30 figs. IX, 296 pp. ISBN 3-540-64730-9

Vol. 237: **Claesson-Welsh, Lena (Ed.):** Vascular Growth Factors and Angiogenesis. 1999. 36 figs. X, 189 pp. ISBN 3-540-64731-7

Vol. 238: **Coffman, Robert L.; Romagnani, Sergio (Eds.):** Redirection of Th1 and Th2 Responses. 1999. 6 figs. IX, 148 pp. ISBN 3-540-65048-2

Vol. 239: **Vogt, Peter K.; Jackson, Andrew O. (Eds.):** Satellites and Defective Viral RNAs. 1999. 39 figs. XVI, 179 pp. ISBN 3-540-65049-0

Vol. 240: **Hammond, John; McGarvey, Peter; Yusibov, Vidadi (Eds.):** Plant Biotechnology. 1999. 12 figs. XII, 196 pp. ISBN 3-540-65104-7

Vol. 241: **Westblom, Tore U.; Czinn, Steven J.; Nedrud, John G. (Eds.):** Gastroduodenal Disease and Helicobacter pylori. 1999. 35 figs. XI, 313 pp. ISBN 3-540-65084-9

Vol. 242: **Hagedorn, Curt H.; Rice, Charles M. (Eds.):** The Hepatitis C Viruses. 2000. 47 figs. IX, 379 pp. ISBN 3-540-65358-9

Vol. 243: **Famulok, Michael; Winnacker, Ernst-L.; Wong, Chi-Huey (Eds.):** Combinatorial Chemistry in Biology. 1999. 48 figs. IX, 189 pp. ISBN 3-540-65704-5

Vol. 244: **Daëron, Marc; Vivier, Eric (Eds.):** Immunoreceptor Tyrosine-Based Inhibition Motifs. 1999. 20 figs. VIII, 179 pp. ISBN 3-540-65789-4

Vol. 245/I: **Justement, Louis B.; Siminovitch, Katherine A. (Eds.):** Signal Transduction and the Coordination of B Lymphocyte Development and Function I. 2000. 22 figs. XVI, 274 pp. ISBN 3-540-66002-X

Vol. 245/II: **Justement, Louis B.; Siminovitch, Katherine A. (Eds.):** Signal Transduction on the Coordination of B Lymphocyte Development and Function II. 2000. 13 figs. XV, 172 pp. ISBN 3-540-66003-8

Vol. 246: **Melchers, Fritz; Potter, Michael (Eds.):** Mechanisms of B Cell Neoplasia 1998. 1999. 111 figs. XXIX, 415 pp. ISBN 3-540-65759-2

Vol. 247: **Wagner, Hermann (Ed.):** Immunobiology of Bacterial CpG-DNA. 2000. 34 figs. IX, 246 pp. ISBN 3-540-66400-9

Vol. 248: **du Pasquier, Louis; Litman, Gary W. (Eds.):** Origin and Evolution of the Vertebrate Immune System. 2000. 81 figs. IX, 324 pp. ISBN 3-540-66414-9

Vol. 249: **Jones, Peter A.; Vogt, Peter K. (Eds.):** DNA Methylation and Cancer. 2000. 16 figs. IX, 169 pp. ISBN 3-540-66608-7

Vol. 250: **Aktories, Klaus; Wilkins, Tracy, D. (Eds.):** Clostridium difficile. 2000. 20 figs. IX, 143 pp. ISBN 3-540-67291-5

Vol. 251: **Melchers, Fritz (Ed.):** Lymphoid Organogenesis. 2000. 62 figs. XII, 215 pp. ISBN 3-540-67569-8

Vol. 252: **Potter, Michael; Melchers, Fritz (Eds.):** B1 Lymphocytes in B Cell Neoplasia. 2000. XIII, 326 pp. ISBN 3-540-67567-1

Vol. 253: **Gosztonyi, Georg (Ed.):** The Mechanisms of Neuronal Damage in Virus Infections of the Nervous System. 2001. approx. XVI, 270 pp. ISBN 3-540-67617-1

Vol. 254: **Privalsky, Martin L. (Ed.):** Transcriptional Corepressors. 2001. 25 figs. XIV, 190 pp. ISBN 3-540-67569-8

Vol. 255: **Hirai, Kanji (Ed.):** Marek's Disease. 2001. 22 figs. XII, 294 pp. ISBN 3-540-67798-4

Vol. 256: **Schmaljohn, Connie S.; Nichol, Stuart T. (Eds.):** Hantaviruses. 2001, 24 figs. XI, 196 pp. ISBN 3-540-41045-7

Vol. 257: **van der Goot, Gisou (Ed.):** Pore-Forming Toxins, 2001. 19 figs. IX, 166 pp. ISBN 3-540-41386-3

Vol. 258: **Takada, Kenzo (Ed.):** Epstein-Barr Virus and Human Cancer. 2001. 38 figs. IX, 233 pp. ISBN 3-540-41506-8

Vol. 259: **Hauber, Joachim, Vogt, Peter K. (Eds.):** Nuclear Export of Viral RNAs. 2001. 19 figs. IX, 131 pp. ISBN 3-540-41278-6

Vol. 260: **Burton, Didier R. (Ed.):** Antibodies in Viral Infection. 2001. 51 figs. IX, 309 pp. ISBN 3-540-41611-0

Vol. 261: **Trono, Didier (Ed.):** Lentiviral Vectors. 2002. 32 figs. X, 258 pp. ISBN 3-540-42190-4

Vol. 262: **Oldstone, Michael B.A. (Ed.):** Arenaviruses I. 2002, 30 figs. XVIII, 197 pp. ISBN 3-540-42244-7

Vol. 263: **Oldstone, Michael B. A. (Ed.):** Arenaviruses II. 2002, 49 figs. XVIII, 268 pp. ISBN 3-540-42705-8

Vol. 264/I: **Hacker, Jörg; Kaper, James B. (Eds.):** Pathogenicity Islands and the Evolution of Microbes. 2002. 34 figs. XVIII, 232 pp. ISBN 3-540-42681-7

Vol. 264/II: **Hacker, Jörg; Kaper, James B. (Eds.):** Pathogenicity Islands and the Evolution of Microbes. 2002. 24 figs. XVIII, 228 pp. ISBN 3-540-42682-5

Vol. 265: **Dietzschold, Bernhard; Richt, Jürgen A. (Eds.):** Protective and Pathological Immune Responses in the CNS. 2002. 21 figs. X, 278 pp. ISBN 3-540-42668-X

Vol. 266: **Cooper, Koproski (Eds.):** The Interface Between Innate and Acquired Immunity, 2002, 15 figs. XIV, 116 pp. ISBN 3-540-42894-1

Vol. 267: **Mackenzie, John S.; Barrett, Alan D. T.; Deubel, Vincent (Eds.):** Japanese Encephalitis and West Nile Viruses. 2002. 66 figs. X, 418 pp. ISBN 3-540-42783-X

Vol. 268: **Zwickl, Peter; Baumeister, Wolfgang (Eds.):** The Proteasome-Ubiquitin Protein Degradation Pathway. 2002, 17 figs. X, 213 pp. ISBN 3-540-43096-2

Vol. 269: **Koszinowski, Ulrich H.; Hengel, Hartmut (Eds.):** Viral Proteins Counteracting Host Defenses. 2002, 47 figs. XII, 325 pp. ISBN 3-540-43261-2

Vol. 270: **Beutler, Bruce; Wagner, Hermann (Eds.):** Toll-Like Receptor Family Members and Their Ligands. 2002, 31 figs. X, 192 pp. ISBN 3-540-43560-3

Vol. 271: **Koehler, Theresa M. (Ed.):** Anthrax. 2002, 14 figs. X, 169 pp. ISBN 3-540-43497-6

Vol. 272: **Doerfler, Walter; Böhm, Petra (Eds.):** Adenoviruses: Model and Vectors in Virus-Host Interactions. Virion and Structure, Viral Replication, Host Cell Interactions. 2003, 63 figs., approx. 280 pp. ISBN 3-540-00154-9

Vol. 273: **Doerfler, Walter; Böhm, Petra (Eds.):** Adenoviruses: Model and Vectors in Virus-Host Interactions. Immune System, Oncogenesis, Gene Therapy. 2004, 35 figs., approx. 280 pp. ISBN 3-540-06851-1

Vol. 274: **Workman, Jerry L. (Ed.):** Protein Complexes that Modify Chromatin. 2003, 38 figs., XII, 296 pp. ISBN 3-540-44208-1

Vol. 275: **Fan, Hung (Ed.):** Jaagsiekte Sheep Retrovirus and Lung Cancer. 2003, 63 figs., XII, 252 pp. ISBN 3-540-44096-3

Vol. 276: **Steinkasserer, Alexander (Ed.):** Dendritic Cells and Virus Infection. 2003, 24 figs., X, 296 pp. ISBN 3-540-44290-1

Vol. 277: **Rethwilm, Axel (Ed.):** Foamy Viruses. 2003, 40 figs., X, 214 pp. ISBN 3-540-44388-6

Vol. 278: **Salomon, Daniel R.; Wilson, Carolyn (Eds.):** Xenotransplantation. 2003, 22 figs., IX, 254 pp. ISBN 3-540-00210-3

Vol. 279: **Thomas, George; Sabatini, David; Hall, Michael N. (Eds.):** TOR. 2004, 49 figs., X, 364 pp. ISBN 3-540-00534-X

MIX
Papier aus verantwortungsvollen Quellen
Paper from responsible sources
FSC® C105338

If you have any concerns about our products,
you can contact us on
ProductSafety@springernature.com

In case Publisher is established outside the EU,
the EU authorized representative is:
Springer Nature Customer Service Center GmbH
Europaplatz 3, 69115 Heidelberg, Germany

Printed by Libri Plureos GmbH
in Hamburg, Germany